Global Asian City

RGS-IBG Book Series

For further information about the series and a full list of published and forthcoming titles please visit www.rgsbookseries.com

Published

Global Asian City

Migration, Desire and the Politics of Encounter in 21st Century Seoul

Francis L. Collins

WILEY Blackwell

Registered Office(s)
John Wiley & Sons, Inc., 111 River Street, Hoboken, NJ 07030, USA
John Wiley & Sons Ltd, The Atrium, Southern Gate, Chichester, West Sussex, PO19 8SQ, UK

Editorial Office
9600 Garsington Road, Oxford, OX4 2DQ, UK

For details of our global editorial offices, customer services, and more information about Wiley products visit us at www.wiley.com.

Wiley also publishes its books in a variety of electronic formats and by print-on-demand. Some content that appears in standard print versions of this book may not be available in other formats.

Library of Congress Cataloging-in-Publication Data

Name: Collins, Francis L. (Francis Leo), author.
Title: Global Asian city : migration, desire and the politics of encounter in 21st century Seoul / by Francis L. Collins.
Description: Hoboken, NJ, USA : Wiley, 2018. | Series: RGS-IBG book series | Includes bibliographical references and index. |
Identifiers: LCCN 2017057649 (print) | LCCN 2017057776 (ebook) | ISBN 9781119380023 (pdf) | ISBN 9781119380047 (epub) | ISBN 9781119379980 (cloth) | ISBN 9781119380009 (pbk.)
Subjects: LCSH: Seoul (Korea)–Social conditions. | Korea (South)–Emigration and immigration. | Municipal engineering–Korea (South)–Seoul.
Classification: LCC HN730.5.S46 (ebook) | LCC HN730.5.S46 C65 2018 (print) | DDC 306.095195–dc23
LC record available at https://lccn.loc.gov/2017057649

Cover design: Wiley
Cover image: *Damunhwa Gil*/Multicultural Street, Ansan City, South Korea © Francis L. Collins

Set in 10/12pt Plantin by SPi Global, Pondicherry, India

The information, practices and views in this book are those of the author(s) and do not necessarily reflect the opinion of the Royal Geographical Society (with IBG).

Printed in Singapore by C.O.S. Printers Pte Ltd

10 9 8 7 6 5 4 3 2 1

Contents

Series Editor's Preface

The RGS-IBG Book Series only publishes work of the highest international standing. Its emphasis is on distinctive new developments in human and physical geography, although it is also open to contributions from cognate disciplines whose interests overlap with those of geographers. The Series places strong emphasis on theoretically-informed and empirically-strong texts. Reflecting the vibrant and diverse theoretical and empirical agendas that characterise the contemporary discipline, contributions are expected to inform, challenge and stimulate the reader. Overall, the RGS-IBG Book Series seeks to promote scholarly publications that leave an intellectual mark and change the way readers think about particular issues, methods or theories.

For details on how to submit a proposal please visit:
www.rgsbookseries.com

David Featherstone
University of Glasgow, UK

RGS-IBG Book Series Editor

Acknowledgements

Like migration, writing a book is not an individual endeavour: it demands support from family, friends, colleagues, community groups and participants; it is often enhanced by research funding; and it has significance through its impact with others. I am immensely grateful to the many people and organisations that have assisted in the writing of *Global Asian City*.

I am grateful firstly to the participants that have informed this book. The time and insight they have offered is what made this research possible; I only hope that the presentation of these stories does justice to their contribution. The completion of the book was made possible by many researchers and colleagues: Đô Dieu Khuê, Vorarerk Khunthongkum, Jeremiah Magoncia and Viko Zakhary ably assisted with interviews; Gil-Sung Park, Doyoung Song and In-Jin Yoon provided generous support and the Asiatic Research Institute at Korea University provided writing space; the *Globalising Universities and International Student Mobilities* research team including Ho Kong Chong, Brenda Yeoh, Mayumi Ishikawa, Jean-Charles Lagree, Nick Lewis, Eugene Liow, Ai-hsuan Sandra Ma, Gil-Sung Park and Ravinder Sidhu were wonderful to work with. I have also benefited from conversations with Tim Bunnell, Jørgen Carling, Yi'En Cheng, Ward Friesen, Elaine Ho, Shirlena Huang, Sergei Shubin, Peidong Yang, Junjia Ye and Sallie Yea. Thanks also to Dave Featherstone for fantastic editorial support and the reviewers who provided constructive commentary that has enhanced the book.

Financial support was generously provided by the Korea Foundation through Field Research Fellowships in 2009 and 2015, the Asia Research Institute and the Faculty of Arts and Social Sciences at the National University of Singapore and the School of Environment at the University of Auckland.

Finally, a special thank you to the most important people in my life, Molly and SeungHee, who have always been supportive, provided advice and guided me through the challenges of research and writing.

Francis L. Collins

Chapter One
Introduction

On a cold morning in early 2008 I was travelling to the Gwacheon Government Complex to undertake an interview with the Korea Immigration Service. As so often happens, it was a chance meeting on this journey that crystallised neatly the context of this book. I was waiting in *Samgakji*[1] subway station in Seoul and was approached by a tall apparently non-Korean man who spoke to me in a thick Texan accent. The stranger struck up our conversation by enquiring about what I was doing in Seoul and where I was travelling to on this occasion.[2] After I made it clear that I was on a journey to interview officials about labour migration the stranger explained that he owned a small manufacturing operation in Incheon where he hired 'a few Filipinos' because Koreans 'expected too much money'. He added that the changes to labour laws for migrants through the Employment Permit System (EPS) meant that he was considering relocation to the Philippines where he could get 'four workers for the price of one here'. As our relatively one-sided conversation continued he informed me that things might get better if the then recently elected President Lee Myung-bak kept his promises and supported businesses over workers; otherwise, 'everyone' was going to leave. Inserted in this commentary was a quip about the 'Filipino condos' he had built for his workers (converted shipping containers where many migrant workers are housed), and a variety of racist comments about the backwardness of Koreans – defended as 'not racist' because his mother is Korean.

The issues discussed in this conversation are demonstrative of the way in which migration has come to be articulated through a distantiation of migrant lives. For the stranger, migration would appear to be a strategy for capital

Global Asian City: Migration, Desire and the Politics of Encounter in 21st Century Seoul, First Edition. Francis L. Collins.
© 2018 John Wiley & Sons Ltd. Published 2018 by John Wiley & Sons Ltd.

accumulation – his investments in Incheon are an attempt to generate higher profits by employing workers for lower wages. As a mobile subject he is empowered by his American nationality, the business visa in his passport and the economic capital he possesses. His movement through local and transnational space appears to be relatively effortless and generated through individualised desires for capital accumulation. If circumstances do not suit he will simply relocate his business activities to a lower-wage environment. He framed himself as an agentive subject of migration. In contrast, the 'few Filipinos' who work for him may have work visas but may also be undocumented; they have much more limited access to migration and under this stranger's logic face the prospect of chasing capital back to their homeland only to be paid lower wages for probably greater work. Rather than being enabled by forms of desire their migration is framed as an outcome of wage differentials and the force of global capitalism.

These migrations are also articulated unevenly through the urban spaces that different migrants come to inhabit. On the one hand, the processes of labour migration to Seoul often takes shape through a peripheralisation of migrant lives. Migrant mobilities link into work and life in distant parts of the Seoul Metropolitan Region like areas in Incheon where small manufacturing operations continue to have a significant presence. Urban life here is often characterised by precarity – living in ersatz accommodation like converted containers, working long hours often for substandard pay and sometimes subject to abusive or exploitative employment. Mobility appears constrained, not only in migration but also in everyday life in the city. By contrast, for the stranger and indeed myself as a visiting researcher, mobility comes to articulate with urban space in quite different ways. We meet by chance in one of the classic foreigner neighbourhoods in central Seoul, reside in comfortable accommodations during our short visits and without the temporalisations of factory work would appear to be able to direct our mobilities through urban space according to our own desires.

By the mid-2000s migration, and the uneven geographies of these and other migrant lives, was becoming an increasingly taken for granted feature of life in South Korea and especially Seoul and its broader metropolitan region, encompassing Gyeonggi Province and Incheon City (Kim, A.E. 2009). In 2007 the Korea Immigration Service announced with some jubilation that the foreign resident population in South Korea had surpassed one million, and that the country was now entering a 'new era of multiculturalism' (Kim, S. 2009); by 2016 the figure had surpassed two million (Korea Immigration Service 2017). For many in the media, politics and the general public, this represented a considerable departure from a national culture that has over the course of the twentieth century emphasised narratives of ethnic homogeneity and shared lineage (Han 2007). In the space of little over a decade, the presence of foreigners in South Korea had shifted from an interesting novelty to one of the critical issues facing society and its future (Kim, N. 2012). This was nowhere more the case than in Seoul, a city that has been represented as the crucible of indigenous economic development for half a century (Kim & Choe 1997) and is now home to the largest number and diversity of foreign residents in South Korea.

Global Asian City explores the entanglement of migratory processes and metropolitan transformations in contemporary Seoul. It does so through an empirical focus on the migration and urban lives of three categories of migrants who have become a common feature of life in Seoul over the last three decades: 'migrant workers', 'English teachers' and 'international students'. The migrants who people these categories have become significant in Seoul both numerically and also in terms of the role of migration in reconfiguring elements of urban life. In 2016 there were 279,187 people holding work visas through the EPS that governs labour migration in South Korea, 76,040 people on student visas and 15,450 people holding language instructor visas of whom English teachers form over 90%. Ordinarily, these migrant populations are addressed in discrete ways in both policy orthodoxy and migration scholarship within South Korea and internationally. They are seen as low-skilled, (potential) elite and middling respectively, and as a result are assumed to be drawn into migration for quite different reasons and to have distinct roles in urban life. Viewed separately, these migration patterns would appear to reflect quite different dimensions of South Korea's recent political–economic history, from the growing labour shortages of the early 1990s (Kim, W. 2004), the transformation of nationally oriented universities into global institutions (Collins 2014a) and the increasing desire for English as a global *lingua franca* (Park, J.K. 2009). Despite their estrangement in scholarship and policy discourse I argue that these migrants and the precursors of their arrival must be conceived concurrently. Their presence, and indeed their socio-political position in Seoul, is very much entangled in processes of national and metropolitan restructuring and in particular the material and imaginative rearticulation of Seoul vis-à-vis national, regional and global assemblages.

This book seeks to bring the narratives that account for these different migrations together and in the process to advance understandings of the relationship between migration and cities. It does so by bringing to the fore the manner that migrants negotiate *both* migration *and* urban life, not as distinct spatial locations and temporal phases of pre-migration, migration and settlement but as always interlinked experiences. Focusing on the conceptual vocabulary of desire, assemblage and encounter, the key claim asserted here is that the urban is the spatial formation through which forms of migration are assembled but also drawn apart and made distinct. Urban spaces clearly play an important role in organising different forms of labour and their linkages into different categories of migrants that are established in the regimes that seek to govern migration. At the same time, the spaces, practices and subjectivities of migrants also need to be examined in terms of the active processes of desiring involved in migration, of seeking better futures, exploring alternative or unknown possibilities and transformations in one's position in the world.

In this opening chapter I set the scene for this contribution by first discussing the recent growth in scholarship on the relationship between migration and cities.

Emerging within both geography and other social science disciplines this literature has advanced beyond a conception of migration as simply an addition of people to cities through a focus on pathways to incorporation, built environment changes and transnational linkages. Yet, the case I make is that there remains either a *migration-centric* or an *urban-centric* outlook in this scholarship where cities remain largely as a backdrop for migration, or the urban lives of migrants in cities are delinked from the generation and governmentality of migration itself. After exploring the geographical and historical backdrop of migration in Seoul and South Korea, the chapter then moves to provide a brief introduction to the conceptual vocabulary of desire, assemblage and encounter and its significance for studying migration and cities. Last, I introduce the research that informs this book, address the analytical challenges and potential of researching migration and cities through different experiences and outline the structure of the chapters that follow.

1.1   Migration and Cities

This book is about the relationship between different forms of migration and the making and transformation of cities. This is not a new concern for geographers or for social scientists. Indeed, the relationship between migration and cities is apparent in urban scholarship dating right back to the work of Robert Park and colleagues in Chicago who traced the arrival, settlement and succession patterns of migrants as part of their primary focus on 'The Growth of the City' (Park, Burgess & McKenzie 1925). Migration was understood as a process of populating cities, a pattern that has been observed in the role of international migration in the emergence of cities such as Chicago as well as processes of internal migration as part of urbanisation, that can be observed in rapidly growing cities throughout the world (McGee 1971). *International* migration has also often been observed for its impacts in specific parts of cities – the manifestations of ethnic enclaves (Portes & Jensen 1989), precincts (Rath 2007) or ethnoburbs (Li 1998) that capture a sense of not only additions to but also changes in the character of urban space. And, migration has formed an important part of arguments about international divisions of labour and socio-spatial polarisation that have been so central in claims about the emergence of global cities (Sassen 2001).

While migration has long been recognised as having a relationship with cities it is less clear that scholars have focused on the specific components of this relationship. Glick Schiller and Çağlar (2009, 2011) argue that this has resulted from a lack of cross-fertilisation between the fields of migration and urban studies. In migration scholarship, for example, 'there are many studies of migration *to* cities and the life of migrants *in* cities but very little about the relationship of migrants *and* cities', while in urban studies migrants appear as members of communities and labour markets but not as key actors in city-making (Glick Schiller & Çağlar

2011: 2). Put another way, there is a tendency for either migration-centric or urban-centric scholarship where only one side of this pairing is properly examined. There is ample literature that sees the city as a backdrop to migrant lives, for example, but does not consider how urban environments are actually reconfigured in the process. Similarly, Chicago scholars and global city theorists alike have focused primarily on what happens after migrants come to the city, not the process of migration itself or its implications in people's urban lives.

Other attempts to explore migration and cities have advanced more focused conceptualisations of this relationship that draw attention to the positioning of cities and the implication this has for migration processes and experiences. Glick Schiller and Çağlar (2009, 2011), for example, have proposed a focus on varying pathways of 'urban incorporation' as key to exploring differences between cities and the varying ways in which migrants become part of urban life. Such an approach involves focusing not only on 'individual migrants, [but also] the networks they form and the social fields that are created by their networks' (2009: 179–180). Accordingly, migrants become 'incorporated' into urban life through different 'pathways' – work, neighbourhood, political and religious organisations for example. The availability of these pathways will differ depending on histories of migration and the 'varying position of cities within global fields of power' (Glick Schiller & Çağlar 2009: 178). Another similar set of arguments has been offered by geographers Price and Benton-Short (2008) who make a case for re-examining the 'immigrant gateway city' concept in a manner that can address the dynamics of contemporary urban life. Here cities are interpreted as 'critical entry points, nodes of collection and dispersion of goods and information, highly segregated settings, sites of global cultural exchange, turnstiles for other destinations, and immigrant destinations and settlements' (Price & Benton-Short 2008: 31). In both propositions, there is a clear sense of the ways in which cities influence the directionality and form of migration as well as scope to consider the broad implications of different types of migration on urban life.

Another set of contributions focuses on the role of migration in urban restructuring and its effects in the built environment of cities. Mitchell's (2004) and Ley's (2010) studies of transformations in Vancouver through migration from Hong Kong are indicative of this genre. Focused on discourse, neo-liberalism and built form (Mitchell 2004) or the unevenness of migrants' economic place in the city (Ley 2010) these studies show the ways in which Vancouver's turn to the Pacific Rim involved not only the arrival of new migrants but also significant implications for the lived experience of urban spaces. Migration was associated with political rationalities of globalisation, economic success and development that would reconfigure the city as a safe space for footloose entrepreneurs. Migration also brought changes however, in the redevelopment of inner city areas, skyrocketing property prices and their association with migration and fear of 'monster' houses changing the character of 'traditional' neighbourhoods. Smith (2000: 5) who has provided his own account of migration, globalisation and built form in Los Angeles captures some of this dynamic in the notion of *Transnational Urbanism*:

> [A] marker of the criss-crossing transnational circuits of communication and cross-cutting local, translocal, and transnational social practices that 'come together' in particular places at particular times and enter into the contested politics of place-making, the social construction of power differentials, and the making of individual, group, national, and transnational identities, and their corresponding fields of difference.

Like Ley (2010) and Mitchell (2004), Smith offers a sense of the transnational dimensions of urban life and in particular the ways in which migration can link together different places and can shape the form and experience of cities.

Lastly, there are studies concerned with the co-presence of migrants and other residents in cities and practices and policies of urban diversity and inclusion. Much has been made, for example, of Vertovec's (2007: 1025) focus on 'super-diversity' as a marker of overlap between ethno-national differences and 'divergent labour market experiences, discrete gender and age profiles, patterns of spatial distribution, and mixed local area responses by service providers and residents'. Migration's diversification in relation to contemporary immigration controls demands a renewed focus on class, gender, race and other axes of social difference in the city (Ye 2016a) that potentially alters presumptions about the pathways of urban incorporation available to migrants (Grzymala-Kazlowska & Phillimore 2017). For Hall (2015), such urban multiculture also provides scope to explore more precisely the reconfiguration or indeed making of cities through the quotidian practices of migrants as ordinary urban residents (see also Collins 2012). In such 'migrant urbanisms' resides scope to move beyond the framing of migrants as particular kinds of ethnic others and to see the daily lives of people on the move as part of the reconfiguration of places rather than only additions to what already exists.

The approach developed in this book contributes to this growing interest in the relationship between cities and migration by foregrounding a conceptualisation of *both* migration *and* urbanisation or urban lives that links the drivers of migration to the different positions that migrants hold in society and the contested politics of everyday life. While the focus on urban multiculture, particularly in its quotidian manifestations, offers scope to examine the everyday constitution of space and the role of migrants as urban actors this is often disconnected from the very conditions that shape migration. In part, this limitation has emerged because research on migration and cities has focused primarily on western immigrant cities as a site for theory making (Collins 2012). In these urban contexts, scholars often take the drivers of migration as obvious (economic advancement, lifestyle, settlement and citizenship) because migration itself is so well established as part of the peopling of settler societies and cities. The experience, direction and implications of migration cannot be taken for granted however, and there is a need to examine how the imaginations of migrants, their desires and aspirations in migrating and the infrastructures that support their movements also reach into the daily constitution of urban life. Moreover, migration cannot be read as a flat experience of similar forms of mobility but rather there is also a need to focus closely on the different statuses accorded to migrants, the temporary forms of migrant entry that predominate in many parts of the world and how its regulation shapes the urban lives of migrants.

Figuring the drivers of migration alongside the varying conditions under which migration takes place is particularly critical to exploring the relationship of migration and cities in Asian contexts. This is not least because of the way that recent patterns of migration in Asia have been tied to the development and globalisation of cities like Hong Kong, Seoul, Singapore, Taipei and Tokyo that have rapidly become important nodes in migratory circuits (Collins 2012; Wong & Rigg 2010). Aside from Singapore, these are not cities with extant histories of international immigration or places that have been long established as desirable destinations for migrant mobilities. What then are the driving forces of migration in this context? How do states seek to manage and modulate migration? And what are the implications for the everyday lives of migrants and other residents in these cities? *Global Asian City* addresses these questions by focusing on the ways in which different types of migration have become viewed as an indispensable dimension of twenty-first-century Asian urbanisation (Battistella 2014; Lai et al. 2013; Ong 2007; Ye 2016a). Migration in many Asian contexts also poses fundamental challenges to extant modes of social and political life, particularly in nations like South Korea and Japan where notions of citizenship are built on seemingly inherent entanglements of race and nationality (Han 2015). Accordingly, the governmentality of migration in Asia has operated through forms of differential inclusion (Mezzadra & Neilson 2013), incorporating migrants as workers, students and spouses but minimizing or eliminating possibilities for other kinds of social and economic interpenetration (Lindquist, Xiang & Yeoh 2012; Seol & Skrentny 2009). *Global Asian City* focuses on the differences established between migrant types; as workers, professionals and students, the way in which they work through the desires migrants express to be mobile as well as shape the lives they can live in migration and the city.

1.2   Migration and Modernity in Global City Seoul

Migration needs to be situated in relation to the varying position of cities vis-à-vis national, regional and global reconfigurations (Glick Schiller & Çağlar 2009). In the case of Seoul, this means accounting for the wider transformation of South Korea in the four decades since the end of the Korean War, a period regularly described as an 'economic miracle' characterised by 'compressed modernisation'

(Chang 1999). Under the authoritarian governments of Park Chung-hee and his successors, the Korean government invested heavily in export oriented industrialisation, encouraged rapid internal migration and enforced strict external controls on outward and inward migration. In the process, South Korean economy, society, culture and politics underwent radical change: a condensed period of industrialisation and economic growth; a subsequent alteration in social and cultural norms manifest most obviously in reshaping of urban life in Seoul and other major cities; tension and competition between familial and societal norms that cut across traditional, modern and seemingly 'postmodern' articulations and a compressed political transition from colonial rule, authoritarian dictatorship to a new political aristocracy operating under the banner of democracy (Chang 1999).

The growth of Seoul was extraordinary during this period as its population increased from 1.6 million in 1955 to 9.7 million in 1985 and in the decades since has sprawled into the wider metropolitan region of Gyeonggi Province and Incheon City where 25.5 million people live, nearly 50% of the country's population. The city already held significant imaginative potential through its history as the Joseon Dynasty (1394–1910) capital of Hanyang and Japanese colonial (1910–1945) capital of Kyeongseong. But industrialisation and enhanced circulation of overseas culture fundamentally reshaped imaginings of the city during this period (Kim Watson 2011). As Jo (2015: 89) notes in her account of poverty and shame in twentieth-century South Korea, 'the city of Seoul was presented as a place of opportunity and hope, embodying a sense of zeal and the heartfelt aspirations of people for a better future'. The city generated a desire for migration and its possibilities that drew at its peak over 300,000 people to the city annually during the 1970s. Unsurprisingly, urban and national governments struggled to cope with this growth: the housing supply ratio reached a low of 53 percent by the mid-1980s and concerns around sanitation, access to toilets and pollution from heating were widespread (Gelézeau 2008). In an indicative precursor to contemporary migration patterns, the reality of everyday life, especially for the large numbers of working migrants in Seoul, rarely met the idealised excitement of a modern metropolis but was rather often articulated through struggles to survive in the uneven and fragmentary spaces of the city. Indeed, as Kim W.B. (1999: 13) put it in his millennial reflection on Korean urbanisation, 'cities were bases for production, rather than places of living.'

If the reconfiguration of Seoul in the mid-twentieth century involved economic, social and cultural extensions into the Korean countryside, then its articulation into the early twenty-first century articulated a transnational augmentation (Moon 2000). By the late 1980s and early 1990s urban challenges of housing, sanitation and heating were beginning to be resolved and the economy was shift-ing from intensive manufacturing towards high-technology supported by increasing levels of education amongst youth cohorts. The 1986 Asian Games and 1988 Olympic Games drew South Korea socially and culturally into a wider set of connections through increased tourist circulation and when emigration controls were lifted in 1989 also enhanced possibilities for travel and migration. Globalisation also emerged on the national political agenda, captured most evocatively by former President Kim Young-sam's flagship policy of *Segyehwa* (literally globalisation):

> Fellow citizens: Globalization is the shortcut which will lead us to building a first-class country in the 21st century. This is why I revealed my plan for globalization and the government has concentrated all its energy in forging ahead with it. It is aimed at realizing globalization in all sectors – politics, foreign affairs, economy, society, education, culture and sports. To this end it is necessary to enhance our viewpoints, way of thinking, system and practices to the world-class level. ...We have no choice other than this (Kim, S.S. 2000: 1)

Segyehwa was a nationalist project of securing the South Korean nation in the future in ways that seem paradoxical with many of the dominant narratives of globalization (Elden 2005). Rather than seeking to alter the social and cultural fabric of the nation, the policies of *segyehwa* have been first and foremost economic and when they have moved into social and cultural areas the focus has more often been on 'upgrading' Koreans' capacity to operate in a wider world rather than dismantling the national project itself (Kim, S.S. 2000; Koo 2007). Globalisation, in this respect, was configured around potential deterritorialisations of economy, society and culture but always with an eye to shoring up the nation, to reterritorialising or stabilising the Korean nation as a transnationally distributed but nonetheless coherent arrangement.

It is also at this time that *international* migration first emerged as a feature of life in Seoul and South Korea more generally in ways that link to the lives of present migrant workers, English teachers and international students. By the late 1980s, many smaller firms, particularly in labour-intensive sectors like garment manufacturing, were facing widespread labour shortages that related to wage increases demanded by an empowered union movement and increased automation and transnationalisation of production in large corporations in particular (Kim, W. 2004; Park, W.W. 2002). Small and medium sized firms in the '3D sectors' (difficult, dirty and dangerous) faced an increasing shortage of Korean workers who were willing to accept the wages or conditions that had previously made these firms internationally competitive (Lim 2003). Transnational migrant labour, initially arriving undocumented but eventually regulated in different ways, provided the solution, a pool of labour that remained cheap and was not subject to the same regulations and rights as Korean workers.

Despite the absence of a labour-importing scheme, records suggested that there were some 6,409 migrants working in South Korea in 1987, many of whom were likely to have arrived on the relaxed tourist visas created for the Asian and Olympic Games (Seol 2000). The number of migrant workers would grow considerably over the following years, to 14,610 in 1989, 21,235 in 1990 and 45,449 in 1991. In the early years, these migrants came from a small number of South Korea's Asian neighbours; China (particularly Korean-Chinese), Vietnam, Philippines, Indonesia and Bangladesh (Kim, A.E. 2009). The government responded through forced repatriation and penalties for employers but they also moved towards providing legal avenues for labour migration through the estab-lishment of the Industrial and Technical Training Program (ITTP) in 1991. Based on the Japanese migration regime, the ITTP allowed for migrants to enter as 'trainees' and then effectively be employed as ordinary workers but without any of the associated legal rights. Over the course of the next decade, the number of labour migrants would increase to 384,000 by 2002 including some 189,000 undocumented (Lim 2002). Under the reformed EPS that will be discussed in Chapter 3 as well as the revised 'Visit and Employment' scheme for Korean-Chinese, these numbers have continued to increase to over 500,000 combined since 2010, including at least 50,000 undocumented workers according to official sources.

It is also over this same period that the number of English teachers began to increase as South Korea became more visible globally and as part of an increasing emphasis on forms of 'global' education signalled in the rhetoric of *segyehwa* (Park, J.S.Y. 2009). While English has been taught in Korea since at least the late nineteenth century it is only since the late 1980s that both Korean businesses and the state have viewed the acquisition of English as crucial to social and economic success. As larger Korean firms began to shift to less labour-intensive, more ser-vice oriented and high-tech activities they have increasingly viewed English as an essential skill in targeting export markets and engaging with foreign companies (Collins & Pak 2008). Essentially, for many large private sector companies 'English is taken to be a sign that the worker is well positioned within the modern world and worthy of a company that aspires to expand globally' (Shim & Park 2008: 148). As a form of 'cultural capital' for operating in a global world the learning of English clearly demands conversational competency that is associated with exposure to 'native' forms.

The focus on communicative competence amplified the emphasis on conver-sational ability and English-only approaches to the classroom. 'Native speakers' became idealised as the best teachers of English, particularly in relation to Korean English teachers who may have been experts in linguistics but were often unable to teach English through English language itself. A significant private English education industry started to grow during the 1990s involving the recruitment of university graduates from several predetermined 'western' nations. In 1985 there were only around 600 individuals on the 'teaching and research' visa that covered language instruction and other education based migration at that time. The num-bers remained low until the mid-1990s, reaching 1,136 in 1993 on the new E2 'language instructor' visa, 4,230 in 1995 and 7,607 in 1997, before declining heavily in the wake of the financial crisis and then reaching 6,414 by 2000. The growth in the number of foreign language instructors increased steadily during the 2000s, to reach 12,439 in 2005 and 23,317 in 2010. This resulted not least from government initiatives like the English Program in Korea (EPIK), and the provincial Gyeonggi English Program in Korea (GEPIK) that sought to place native speakers in the public schooling system. Since 2010 there have been incremental declines in the number of E2 visa holders to the current number of 15,450 in 2016.

More recently, this emphasis on educating 'global subjects' (Kang 2012) has also become incorporated into the restructuring of universities as 'world class institutions', through an increasing emphasis on research output and rankings and the related attraction and retention of international students. International students are particularly valued for the role they play in transforming campus spaces, and in providing Korean students in domestic institutions with exposure to a wider range of cultural and linguistic forms (Moon 2016). Moreover, international students are also being viewed as future skilled labour in training, and their presence has been conceived in terms of their potential role in 'future international business and trade relations' (Kim, S.K. 2013). Many international students originate from parts of Asia and their cross-cultural skills and training is claimed to open opportunities for them in an increasingly transnationalised Korean economic sphere (Shin & Choi 2015). In contrast to migrant workers, then, who are positioned at the bottom of labour market hierarchies, and English teachers who occupy a unique middling niche, (graduating) international students are often construed as key contributors to the present and future transnationalisation of Korean political–economic life.

Unlike both migrant workers and English teachers, there remained relatively few international students in South Korea until the early 2000s. In 1985, for example, there were only 433 people on study visas, the largest number of whom were from the USA and Japan, reflecting the role of exchange and area studies under Cold War geopolitical settings. In 1990 the number had increased to 803 and then to 1,487 in 1995 and started to include a wider range of nationalities, including Malaysians who represented about 8% of all students. It is not, however, until the following decade that international student numbers increased considerably and started to follow the current nationality trends. In 2000 there were 3,762 students, 20,683 in 2005, and then increasing quite rapidly to 69,600 by 2010. International students are now also much more likely to come from within Asia than they did in the past. Some 95% of all international students are of an Asian nationality, with about 40% coming from China and significant numbers of other students from Mongolia, Vietnam, Japan, Indonesia and India (Korea Immigration Service 2017).

The growing number of migrant workers, English teachers and international students in Seoul and South Korea over the last few decades reveal the manner in which the transformation of this city has involved an extension into previously unfamiliar territories of economic activity, educational forms and social and cultural possibilities. They are particularly indicative of the manner in which the city of Seoul stakes a claim to be global, signalling the ongoing robustness of manufacturing, the linguistic and educational transformation of young people and a new role as a global centre of knowledge production and circulation. Other migration streams that are beyond the scope of this book and the research that underpins it have become significant during this period as well. Most notable has been the growing number of international 'marriage migrants' in South Korea, particularly women from other parts of Asia marrying Korean men. Often established through brokered relationships and directed at least initially towards rural areas, marriage migration has been prominent in public, political and scholarly discussions of migration and multiculturalism in South Korea in recent decades (Kim, A.E. 2009; Kim, M. 2013; Lee 2008). This is not least because of the role that migrating women have been expected to play as mothers of the next generation of Koreans in a context of ultra-low fertility. While this text does not extend to examining these lives, they are significant for apprehending the reconfiguration of the Korean nation around migration. Indeed, as will become clear in later chapters there are instances where marriage migrants transform into independent labour migrants and vice versa, and the discourses of multiculturalism often operate through reference to the centrality of marriage migrants over those who ostensibly come to labour and study.

1.3   Desiring Migration and Urban Encounters

As the foregoing discussion has demonstrated, the city and urban space more broadly represent the critical conjuncture of the different migratory flows that are now occurring through South Korea. Seoul is not simply the 'context', 'entry point' or 'setting' (Price & Benton-Short 2008) of *migration*, the point at which movement stops and a more sedentary life begins, but is rather entangled in the multi-scalar process of mobility, where large-scale movements intersect with smaller ordinary events of everyday life. This is a key claim that is extended throughout this book, that the urban is a spatial formation through which the uneven, or discrepant, experiences of migration are assembled. Urban spaces play a critical role in organising different forms of migration, from the political–economic transformations that manufacture demand for labour to cultural, economic and technological imaginings of the city as a space of desire. At the same time, the variegated landscape of migration, the different rights, tenure, tempo and futures of migrants, emerge most obviously through their place in the city, from prominent roles in the stylish sectors of the urban core to seemingly marginalised roles that represent the 'underbelly' of globalisation (Ong 2007). This distantiation of migrant lives is obviously demarcated first at the border, and through the different modes of entry that migrants take to national space, but as this book will demonstrate, the border cannot be conceived as a singular point or moment but rather as a technology that extends into the everyday lives of migrants and their prospects for incorporation in the city (Mezzadra & Neilson 2013).

The migration regimes that have been established in South Korea over the last three decades play a significant role in constituting a cultural politics of migration that is animated by both aspiration and anxiety. Migration expresses aspirations by nation states to be part of flows of knowledge, capital and actors and a wider national desire for economic development and global emergence. There is then a desire for freedom and circulation in migration regimes that promotes movement between places whether of diasporic returnees, mobile professionals and labour, or international students. This desire, however, also clearly runs in tension with anxieties about order, national security and cultural homogeneity, the deterritorialising effects of migration on local populations and social milieus. The identities of migrants are inscribed in this tension as they traverse and reside in the urban spaces that emerge in articulation with migration regimes. In the process, they are subject to graduated forms of incorporation, they are marked as more or less desirable forms of labour, as amplified symbols of global status or derided and made invisible through their peripheralisation, and differentially positioned socially and geographically in relation to locals and other mobile subjects.

The key argument posited in this book is that in their imaginaries, migration and everyday lives, migrants transcend these migration regimes and their attendant politics of truth. In their very being and becoming in the world, migrants both appropriate and rework the territorialising powers of migration regimes – they become both the labouring and learning bodies desired in these regimes but also active human subjects whose presence can never be completely contained. This reflects the force of desires involved in generating migration and sustaining migrant lives, not just in the initial impetus of departure but also across the spatial and temporal horizons of movement and possibility. Migration is caught up in a 'desire to circulate' (Raghuram 2013) that results from individual relations with collective imaginaries about the value of movement and the attributes of particular destinations. Migration is then necessarily strategic even though it can never be known or completely planned in advance; migrants take opportunities they are presented with and seek to materialise their desires in place while governments actively seek to shape those desires. As this book will demonstrate, whether workers, teachers or students, migrants manoeuvre between the governmental will of migration regimes and their own embodied desires to perform, transcend or escape this will. They become in circulation in ways that cannot be fully circumscribed but also are never completely knowable to themselves in advance.

In the discussions that follow in this book I take up this tension between constraint and possibility in migration and urban life by linking ideas of desire and assemblage with a focus on the encounters that constitute urban life. This framework extends current understandings of migration and cities by addressing how the generation of migration (desire) occurs across the mobilities and lives of migrants, is shaped by shifting urban, national and global forms and regulation (assemblage) and has consequent impacts on the everyday lives and encounters of migrants in the city. In empirical terms, this approach draws attention to

Seoul's present reliance on migrants and the way in which it is situated in its histories and geographies of modernisation and globalisation; how shifting imaginative geographies of Seoul and South Korea are generating new forms of migration that are marked by considerable spatial, social, legal and economic divisions; and how the arrival of diverse migrant populations is altering not only the fabric of particular parts of Seoul but also its prospects for a globally oriented future. Put theoretically, I read migrant mobilities through their biographical background, movement across borders and everyday lives; examine the city as an assemblage that has historical depth but also varying spatial extension manifest through these mobilities; and position encounters with urban life and residents as part of the coupling and decoupling of migrant and urban futures.

This focus on desire, assemblage and encounter makes two interlinked contributions to migration and urban geographies. First, introducing a focus on *desire* as a social force provides an alternative conceptualisation to the still common reading of migrants as utility-maximising individuals who, when provided with full access to information, can make migration decisions that will serve their own interests in a goal-oriented manner (Nail 2015). While migration scholars rarely explicitly advocate this principle there has been little theoretical advancement of the drivers of migration (Carling & Schewel 2017) and in the absence of alternative conceptualisations 'the utility-maximizing notion underlying decision-making has not been fundamentally challenged' (De Haas 2011: 20). Scholarship on migration and cities has effectively taken this utility-maximising model for granted by focusing discretely on the role of cities in migratory patterns or the everyday lives of migrants rather than seeing these as interlinked from the outset. In contrast, I advocate a reading of desire as a social force as the basis of an alternative conceptualisation of migration that goes beyond the utility-maximising approach and links the generation of migration with differential incorporation into urban life.

Inspired by the work of Deleuze and Guattari (1983, 1986), desire is conceived here as the energies that draw different bodies (human, non-human, symbolic) into relation with each other and in the process leads to shifting social and material formations, such as cities. Desire is evident in the way that migration occurs through strategic planning, opportunism and fancy that lead individuals to move locally, nationally or internationally, to achieve or avoid (un)desirable futures. The focus on desire as a more embodied and socially connected force is important for migration geographies for three reasons: (1) it displaces the presumption that migration results only from calculated economic rationalities; (2) it reveals the way that seemingly individual 'decisions' are generated in wider imaginative and social circulations; and (3) it demonstrates that what we call 'decision-making' does not occur at a single point in time but is rather stretched across the migration individuals undertake and is constantly reconfigured through encounters across and within places. In *Global Asian City*, this focus on desire makes it possible to conceive of what appear to be very different forms of migration – workers, teachers and students – as part of similar processes of encountering the idea of migration and the possibility that exists in Seoul as an emerging global city. Rather than viewing these migrations as completely incommensurable at the outset, the focus on desire shows that it is the encounter with the globalising urban assemblage of Seoul, and more broadly the nation state of South Korea, that amplify discrepancies in these migrations and their effects in everyday life.

Second, then, *Global Asian City* also extends recent efforts to conceive of cities as urban *assemblages* by drawing attention to the importance of migration in the transnational extension of urban connections and their territorialisation in everyday life. I argue in particular that urban spaces play a critical role in generating and organising different forms of migration, from the political–economic transformations that manufacture demand for labour to cultural, economic and technological imaginings of the city as a space of desire. At the same time, the variegated landscape of migration, the different rights, tenure and tempo that is generated in the regulation of migration, emerge most obviously through their place in the city, the different politics of encounter that are generated and experienced by migrants. While theorisations of assemblage urbanism (McFarlane 2011a; Edensor 2011) and postcolonial readings of cityness (Simone 2010) remain open to the role of migration, they have not to date explicitly explored the role migration plays in the constitution of urban life. This empirical gap also provides scope to advance theoretical understandings of assemblage because migration reveals how desires emanate from particular assemblages such as cities and nations, and their economic, social and imaginative potential. In *Global Asian City*, for example, the focus on desire reveals how the imaginative geographies of an advanced and desirable Seoul/South Korea circulates widely in Asia to generate the migrations of workers and students, while in the case of English teachers these imaginative geographies typically subsume Seoul and South Korea into wider imaginings of Asia or the East. These expressions of desire matter not only for generating migration, but also for then shaping how individuals become part of urban assemblages in their daily lives – their perception of what Seoul and South Korea should be like, their orientation to individuals they encounter and their visions for their own futures.

The analytical connection between notions of desire and assemblage rests on carefully examining the *encounter* between individual migrants and urban life, the effects of national migration regimes, and the influence this has on the negotiation of everyday spaces. Encounter is not simply the meeting of two separate entities, migration and cities or people from different places, but rather emphasises a process of becoming and change (Amin & Thrift 2002; Simone 2010; Wilson & Darling 2016). Migration is having a fundamental role in altering the present position of Seoul vis-à-vis other territories and in turn migration and migrant lives are also modulated in important ways by their encounter with Seoul and South Korea. In *Global Asian City* this is exemplified by drawing attention to each migrant group's position in the schema of ethnic difference dominant in Korean society, the impact of different visa statuses, the range of encounters that different individual migrants have at work, study and in public life, and the possibilities that exist to create a new place in the city. Often these encounters align with the regulatory distinctions between workers, students and teachers but *Global Asian City* also emphasises how gender, race, age and geography cut across the lives that migrants have in Seoul. Most importantly, *Global Asian City* emphasises that desire expressed in migration and the encounter of migrant lives with the city exceed the actual event of this encounter (Karaman 2012). In this sense, focusing on encounter highlights transformations that emerge in the rhythms of daily life as well as the future possibilities manifest in connections between migration and the city. The narratives in this book reveal the ways individuals negotiate alternative futures as long-term residents or as mobile cosmopolitan subjects. For Seoul too, the encounter with migration implies a reconfiguration of the urban assemblage in ways that remain under examined to date, the emergence of new alliances and subjectivities, the transformation of local spaces and the re-imagination of what it might mean to be a *Global Asian City*.

1.4   Approaching Discrepant Lives

The narratives of migration and urban life presented in *Global Asian City* emerge from fieldwork conducted in Seoul and its surrounding areas between 2008 and 2012 and a subsequent period of six months in 2015. The research comes from two projects that had quite different empirical scope. First, material on migrant workers and English teachers emerges from a project on *Mobility, Social Difference and Urban Incorporation*. This project was explicitly comparative and sought to explore the migration and urban lives of migrant workers and English teachers and their role in urban transformations in the Seoul metropolitan region. It included 40 biographical interviews with migrant workers from four Southeast Asian nations, Indonesia, Philippines, Thailand and Vietnam, which at the time of the research were the four largest nationalities within the EPS. In order to capture the depth of meaning in migrant narratives, the interviews were undertaken with research assistants from the same national backgrounds as migrants and in their native languages. As part of the same project, 41 biographical interviews were undertaken with English teachers from Australia, Canada, New Zealand, South Africa, the UK and USA. The interviews for both groups were guided by a general biographical structure that focused on key points before migration, during life in Seoul and future projections, although the interviews themselves always diverged into wider questions raised by migrants about their own individual trajectories, their subjective transformations and their sense of place in Seoul.

The second body of research material that informs the discussions in this book comes from a collaborative, multinational study of *Globalising Universities and*

International Student Mobilities in East Asia. This project explored the restructuring of nine leading universities in the East Asian region and the narratives of international students moving through these institutions and the cities and nations in which they are positioned. Two of the nine case study universities in this project were Korea University (KU) and Seoul National University (SNU), arguably the most prestigious private and public universities in Korea (the study included seven other universities in China, Japan, Singapore and Taiwan). Within this study 40 biographical interviews were carried out with international students at KU and SNU. I conducted 20 of the 40 interviews in English and in some cases partially in Korean; the remainder were conducted by other team members in Mandarin and English. Although these interviews focused on specific issues related to educational migration they were also guided by a similar biographical structure that included opportunities for participants to develop narratives about their experiences prior to coming to Seoul, their everyday lives both on and off campus, and their future projections. The interviews from both of these projects have been transcribed and where necessary translated into English prior to analysis through an open-coding schema.

The integration of research across different migrant populations and indeed drawing from different research projects clearly raises methodological questions that pertain to the claims that can be made in this book. There is, first, a variation in depth of engagement with different research participants. As noted, I was only involved in some of the interviews with international students and worked to train research assistants to undertake interviews with all of the migrant workers in this study. There were many reasons for this design, including the importance of using languages that participants can confidently communicate through as well as creating research encounters that were meaningful and comfortable. Research assistants were native speakers of participants' languages and in the case of migrant workers were all international students of the same nationalities attending universities in Seoul; in the case of international student interviews it also included other research team members. Research assistants were trained in qualitative methodologies and the ethics and techniques involved in undertaking research on people's lives and during the research process I worked with these researchers to reflect on emergent insights and develop their skills in interviewing.

This approach to working with research assistants in the case of interviews with migrant workers was part of an attempt to develop broader comparable insights into the lives of different migrants. It does contrast with both of the other cases where due to language abilities I was able to conduct many of the interviews. It differed in particular from the research undertaken with English teachers where my own experience as an English teacher in Seoul in the early 2000s has provided additional insight into the configuration of teacher lives that was not immediately available in other cases. In order to address these differences, each of the research assistants was asked to provide research journals of their experiences and reflections that emerged in the interviews with migrant workers.

The interviews were subsequently translated at a later date by other assistants who provided guidance in interpretation of cultural and linguistic differences. Interview narratives were also supplemented by visits to spaces where different migrants socialise that included observation and informal discussions with people holding work and study visas. An additional level of insight comes from interviews with government officials and NGOs and discussions with Korean migration scholars in the case of migrant workers; university officials and student association representatives in the case of international students; and for English teachers, representatives of the Seoul Global Center and an English language publication and leaders within the Association of Teachers of English in Korea. Although I do not draw on these observations, key informant interviews and discussions explicitly in this book they form an important component in developing insights into the differences and similarities involved in the migration and urban lives of migrant workers, English teachers and international students.

While several methods could have been employed to undertake this research, such as ethnographic work, photo-voice, participatory mapping or go-alongs (Collins & Huang 2012), interviews were chosen for two reasons. First, interviews are a valuable technique to explore migratory processes because they encourage participants to construct a narrative around their mobilities, to describe key features of their everyday lives and to articulate visions for the future. As geographers exploring migration have demonstrated (Rogaly 2015), the stories that are told about mobility are critical to contextualizing the territories where movement is articulated as well as providing insight into the temporal unfolding of the biographies of individual migrants. Migrant stories generated through interviews 'can reveal the empirical disjuncture between expectations of migration, produced through dominant and pervasive discourses of modernization, and the actual experiences of migrants' (Lawson 2000: 174). The interview schedules employed in this research, and the training provided to research assistants, emphasised an examination of migrant biographies, tracing lives and aspirations from well before participants came to or even thought about Seoul and across the different encounters and experiences that formed part of their mobility and urban life.

In addition to its analytic value, interviews were also the most pragmatic method, particularly in terms of the time demands on participants who were migrant workers as well as the requirement for comparability across international case studies of international students. Many migrant workers who participated in this research had very little time outside of their workplaces and it was considered unreasonable to expect significant commitment to more detailed research techniques or multiple interviews. As the discussion in Chapter 4 highlights, many migrants work six or seven days a week, often from very early in the morning to late in the evening. Time is at a premium and the day off is a time that is precious, for socialising with friends, relaxing or undertaking non-work pursuits. The biographical interviews used in this research are fit for this purpose. They have served as a valuable technique for generating insight into the lives of migrants and providing scope for their own narrations of mobility and everyday life.

Given the complexity of processes involved in enacting and experiencing migration and its articulation through urban spaces, how can we go about accounting for differences across groups? How might we account for the different but intersecting ways in which migrant workers, English teachers and international students become part of the fabric of urban life in Seoul? In what ways can comparative gestures form a part of critical scholarship without reinforcing ideological difference around migrant status, nationality or skill-level? How can we account for ethnicity and identity without resorting to an ethnic lens that can shroud key issues that cut across migrant lives?

Global Asian City is necessarily framed as a comparative project, one that seeks to draw together the lives of different migrants and to explore their articulation in and through the urban fabric of Seoul. In focusing on these three groups, however, I explicitly seek to eschew orthodox approaches to comparison that tend towards identifications of similarities and differences across migrant groups. The purpose of focusing on these three groups is not to establish universal claims about migration, to establish these three cases as plural and incommensurable or to demonstrate that differences emerge because of internal or inherent qualities of each group. Instead, the comparative focus in *Global Asian City* is guided by more relational terms, an emphasis on the connections that cut across the migration of these groups and the ways in which their seeming categorical differences are actively assembled and reinforced in migration and everyday life.

The focus on examining transnational migrations that are not ordinarily conceived in relation to each other – migrant workers, English teachers and international students – foregrounds the relationship between border crossings and urban lives. Ordinarily, these different mobile subjects would be presented as broadly representing groups perceived as low-skilled, middling and (future) elite, and because of their ostensible differences are often held apart analytically, as components of different types of mobility practices, viewed differently by the state, and at times seen as incommensurable. While the distinction between migrant types has been a powerful tool for migration scholars this analytical strategy also occludes a number of issues that are critical to understanding contemporary transformations in migration. First, accepting the separation of migrants serves to reify the state and its practices of distinction as the most important determinant of migration, where research is increasingly suggesting a more complicated landscape of actors involved in migration regimes (Lindquist, Xiang & Yeoh 2012; Glick Schiller & Salazar 2013; Kalir, Sur & van Schendel 2012). Second, the analytical separation of migrants ignores the obviously relational characteristics of mobility itself, where certain kinds of migration are enabled at the expense of others (Cresswell 2010; Collins 2009). Finally, focusing on distinct types also undermines our capacity to appreciate the power and influence of ideology in shaping differences and similarities between groups (Grossberg 1993).

In linking these three different transnational migrations I posit the urban as the spatial formation through which 'discrepant experiences' of mobility are articulated. This approach builds on Edward Said's classic discussion of discrepant experiences in *Culture and Imperialism* (see also Kim Watson 2011; Rubin 2012). Said's claim is that once we recognise the 'knotted' histories of ostensibly different categories and experiences, there can be no intellectual reason for giving them an idealised and essentially separated status in analysis. Indeed, drawing categories apart analytically is part of the generation of ideologically entrenched difference, as is very much the case in migration where categories are also largely determined by the state for its own ends. Intellectually, we must be able to consider and think through experiences that are 'discrepant', that do not neatly map on to each other and that have their own internal formations, coherence and system of external connections. Juxtaposition is a powerful tool in this respect, that creates possibilities to conceive of the generation of difference:

> to make concurrent those views and experiences that are ideologically and culturally closed to each other, and that attempt to distance or suppress other views and experiences. Far from seeking to reduce the significance of ideology, the exposure and dramatization of discrepancy highlights its cultural importance; this enables us to appreciate its power and understand its continuing influence. (Said 1993: 37)

Said describes this as a 'contrapuntal' rather than comparative perspective. In music, the term contrapuntal is derived from the Latin expression *punctus contra punctum*, which literally refers to 'point against point' or, 'note against note'. The term is commonly used to refer to a combination of simultaneous parts, usually melodies, where each has its own independent significance but together results in a coherent texture (Kennedy et al. 2013). Analytically, then, a contrapuntal perspective encourages us to recognize the role of seemingly independent, or discrepant, phenomena within broader sociospatial arrangements. This is critically different from orthodox approaches to comparison, particularly within migration and urban studies, where the focus has too often been on comparing groups or entities that are conceived as having equivalence. In migration studies this manifests in the prevalence of the 'ethnic lens' in studying migration in national contexts (Glick Schiller & Çağlar 2009), and in urban studies in the tendency to compare cities within particular national contexts, or similar levels of economic development (Robinson 2011). The contrapuntal approach taken here is more akin to ideas of 'relational comparison' that stress a focus on interconnected trajectories rather than 'searching for similarities and differences between two mutually exclusive contexts' (Ward 2010: 480). This approach allows for a focus on the interconnected histories and geographies that are involved in the ongoing (re)production of difference in migration and urban life.

The contrapuntal approach undertaken in *Global Asian City* manifests through discussions that cut across various spatio-temporal dimensions of desire, assemblage and encounter. The next chapter, *Desire, Assemblage, Encounter: Beyond Regimes of Migration Management*, elaborates the theoretical focus on desire, assemblage and encounter in relation to policy orthodoxies of migration management and critical accounts of migration regimes. The aim of the chapter is to establish a conceptual vocabulary for analytically integrating the lives of migrant workers, English teachers and international students and their present and future place in Seoul. The chapter problematises the political rationalities of migration management and their theoretical underpinnings within mainstream migration studies, which at once reduce migrants to utility-maximising individuals while also elevating the state to a pre-eminent role in dictating and directing migratory flows. In contrast, I build on insights developed in the study of migration regimes to reveal some of the wider sets of actors and flows involved in constituting contemporary migration as well as the importance of recentring migration studies on migrant lives. In order to address more specifically the spatial depth of migration across urban, national and transnational territories as well as the questions of agentive will in migrant lives the chapter then moves to introduce in detail the conceptual potential of a focus on desire, assemblage and encounter. As a vocabulary for approaching discrepant migrant lives these ideas provide the tools to reassess the generation of migration beyond economic rationality, the ways in which this is shaped by shifting urban forms and migration regimes and the consequent impacts on the everyday lives and encounters of migrants in the city.

The specific function of migration regimes in relation to the role of desire in migration is taken up in Chapter 3 that focuses on the South Korean nation state and its technologies of bordering and migration management that have come to shape migration and lives of migrants. The focus is on the governmentalities involved in the different technologies for regulating migrant mobilities, the EPS for migrant workers, the English Teacher Program, and the International Student visa system. Critically, I am interested here in how these migration regimes reach across borders and into the spaces and lives of migrants – generating desires for migration but also shaping, channelling and circumscribing those desires through the migration process. In this respect, the chapter demonstrates that across these groups migration regimes include not only sending and receiving nation states, but also a wider range of actors and processes – from families, communities, intermediaries, employers, popular culture and education – that contribute in different ways to the prospects and problems of migration.

The next three chapters shift focus from the generation and management of migration to the articulation of migratory processes in the everyday lives of workers, teachers and students. Each chapter focuses on the ways in which these different migrants are drawn apart in their migration, into different spaces and times in the city, with different opportunities for encounter and incorporation and as a result quite different politics of mobility. Following the contrapuntal approach taken in this book, each chapter prioritises the narratives of one of these ideologically constituted migrant categories while also showing intersections with others. The result is a clear sense of the internal formations, coherence and patterns of migration and urban life for workers, teachers and students, and also sets of external connections and overlaps that reveal the workings of migration regimes and the influence of desire and urban form in constituting discrepancy.

Chapter 4 begins by exploring the relationship between *Migration, the Urban Periphery and the Politics of Migrant Lives*. It draws on Simone's (2010) conceptualisation of the urban periphery as a space of marginalisation but also opportunity and draws attention to the role of this urban space in the lives of migrant workers. The periphery here serves as a material manifestation of marginalisation for many, their distance from the urban core and invisibility as urban residents. Yet, by focusing on forms of irregular migration, the social networks that constitute the 'mobile commons' (Papadopoulos & Tsianos 2013), and the politics of recognition, this chapter demonstrates that the periphery also supports other prospects for migrants that have the potential to transform their place in the city. Chapter 5 focuses on international students and the ways in which their narratives point to a channelling of desiring migration and segmentation of migrant lives in the city. As the chapter demonstrates, students move between highly mediated spaces on campus to wider encounters with Korean publics in the dynamic spaces of the urban core. Their mobility through these spaces and encounters with different publics are notably influenced by configurations of race, gender, Korean language ability and cultural norms. While these configurations empower some students to envision and craft a successful future, others experience dissonance with life in Seoul that disrupts both their own conception of the value of international student mobility and the ostensible valorisation of international students as agents of a globalising society and economy. Last, Chapter 6 focuses on English teachers, whose presence in more ordinary neighbourhoods across Seoul might frame them as a more privileged migrant subject, not least because of the way in which whiteness has come to be idealised in English education and globalised urban life in Seoul. English teachers, however, are often in precarious positions, their lives and social networks shaped by transience and for those who remain long-term in the city their privileged labour market niche can circumscribe their conceptions of what is possible in the present and the future.

These earlier chapters set the stage for raising questions about the materialisation of urban multiculturalism and the prospects for migration in Seoul, South Korea and indeed across East Asia. Chapter 7 expands the temporal and spatial horizons of earlier chapters to examine the ways in which migrant lives take shape in the city through specific sets of social practices and relationships that cut across differences established in the migration regime and take shape in both consolidation and disruption of extant identities. Focusing on *Multicultural Presence and Fractured Futures*' this chapter also addresses the ways in which migrant lives are coupled and decoupled from the making of the *Global Asian City*. The discussion is situated within emerging multicultural discourses in South Korea and their articulation into the politics of everyday life and the role of migration in Seoul's urban future. Workers, teachers and students are unevenly incorporated into these visions for the future. They are all viewed as part of a multicultural presence in the now, but their involvement in the future of Seoul is necessarily shaped by a complex intersection of labour market and migrant status, nationality, ethnicity, gender and language that constitutes their desirability as mobile subjects but also as urban residents. In concluding, I return to the critical concerns of this text around migration, desire and the politics of encounter and consider their implications for conceiving and engaging with migration and urban life in an East Asian region where this nexus is increasingly central. As *Global Asian City* demonstrates, we are forced to recognise an urban future actively being assembled as regionally and globally connected and yet clearly reliant on the uneven inclusion of the mobile subjects who are all critical to enabling this globality.

Acknowledgements

Portions of this chapter have been drawn from Collins, F.L. (2016). Migration, the urban periphery, and the politics of migrant lives. *Antipode*, 48(5), 1167–1186. doi: 10.1111/anti.12255 and Collins, F.L. (2016). Labour and life in the global Asian city: the discrepant mobilities of migrant workers and English teachers in Seoul. *Journal of Ethnic & Migration Studies*, 42(14), 2309–2327.

Endnotes

1   *Samgakji* station is located at the west end of the Yongsan Garrison, a key military instalment for US forces in South Korea and before that the headquarters of the Imperial Japanese Army. The area around Samgakji is then unsurprisingly characterised by a significant number of foreign residents and activities.
2   While such an approach might seem unusual in a mega-city like Seoul, it is relatively common for non-Koreans to approach each other in the street and start conversations. This has certainly been my experience but it is also something noted by many of the participants in this research. It would seem to highlight the relative rarity of encounters with non-Koreans but also perhaps the desire for contact in familiar languages and cultural norms that can sometimes be difficult to establish in migration to a place like Seoul.

Chapter Two
Desire, Assemblage and Encounter: Beyond Regimes of Migration Management

Central to managed migration is the establishment of a regime that is capable of ensuring that movement of people becomes more orderly, predictable and productive, and thus more manageable. Based on the principle of regulated openness and sustained by close cooperation between nations, the new arrangement will avoid knee-jerk reactions to the rising emigration pressure and will seek, instead, to bring emigration pressure and the opportunities for legal and orderly entry into a sustainable harmony. In doing so, it will balance and harmonise the needs and interests of the sending, receiving and transit countries and the migrants themselves (Ghosh 2007: 107).

Contemporary forms of migration are increasingly organised through the logic of 'managed migration' proposed above by Bimal Ghosh. A former senior director and consultant with the United Nations (UN) and the International Organization of Migration (IOM), Ghosh has led calls for the establishment of an international regime for the 'orderly' control of migration across national borders. In contrast to approaches that seek to exclude or actively reduce migration, a focus on migration management hinges on claims about the value of 'regulated openness' where migration occurs under particular conditions and where the shared benefits of migration lead to 'sustainable harmony'. This utopic vision of globally integrated migration has manifested in a range of national policy frameworks, regional agreements and in the actions of international organisations (Geiger & Pécoud 2013). The Employment Permit System (EPS) that is currently used to manage labour migration in South Korea is an apposite case (ILO 2010): it involves

Global Asian City: Migration, Desire and the Politics of Encounter in 21st Century Seoul, First Edition. Francis L. Collins.
© 2018 John Wiley & Sons Ltd. Published 2018 by John Wiley & Sons Ltd.

tri-partite negotiations in South Korea, has been developed through government-to-government agreements with sending countries, guarantees certain minimum rights for migrants, and is based on the principle of temporary migration and return. The claim, echoing Ghosh's argument, is that the 'EPS will serve as a stepping stone to the economic growth of this country and, at the same time, a source of dream and hope for foreign workers' (Min 2011: 5). The migration regimes that govern the movement of English teachers and international students in South Korea similarly prioritise the management of migrants through claims about the mutually beneficial character of migration when it is orderly, predictable and productive (see Chapter 3 for a more detailed discussion).

A close examination of Ghosh's claims about the rationale for and benefits of managed migration reveals the way in which ideas about the appropriateness and character of migration can establish powerful frameworks that reproduce inequality and precarity in migrant lives. Migration, or more precisely 'emigration pressure' is introduced here as a problem in need of management. Rather than recognising migration or mobility more generally as a fundamental tenet of human existence, managing migration implies that appropriate forms of life should be established within relatively fixed territorial spaces and movement across borders should only occur when it is 'orderly, predictable and productive'. By contrast to their diminishment of migration, these rationales for 'management' elevate the nation state and its partners in international organisations as the most important actors in determining the right to and conditions of movement. Their task is to control movement, to determine who can move and under what circumstances, how long they can remain and to reconcile these arrangements with an understanding of the nation as the sacrosanct socio-geography of human life. As Nail (2015: 5) puts it more generally in relation to conceptions of migration and movement, 'the migrant [is] understood as a figure without its own history and social force'. Rather, the histories, presents and futures of the migrant have been scripted through the lens of the nation state and citizen where stasis rather than movement is taken as the normal state of being (Casas-Cortes et al. 2015).

These political rationalities of migration management operate through a view that migration occurs either as a singularly economic undertaking or as a result of forced movement. Migrants are either pursuing specific and calculable economic objectives or are stripped of agency because their migration is completely forced or involuntary (Erdal & Oeppen 2017). This understanding of migration constitutes a powerful example of what Pécoud (2014: 3) calls 'international migration narratives … a relatively coherent body of knowledge and ideas, regarding both *what migration is* (trends, numbers, dynamics, etc.) and *what it should be* (through the elaboration of so-called policy recommendations)'. Such narratives, as in the case of migration management, are the product of a growing corpus of knowledge about migration generated in government policy and analysis, the reports of international organisations and in the work of academics studying migration. At the core of much of this knowledge is a particular figure of the migrant as a utility-maximising individual who – when provided with full access to information and freed from constraints – can make migration decisions that will serve her own interests and for which she can be held responsible (De Haas 2011). While the work of international bureaucrats and national policy makers can be identified as part of the production of this figure of the migrant, it is also clear that migration studies itself has often replicated and contributed to these understandings (Garelli & Tazzioli 2013).

These understandings of migration manifest most obviously in the frameworks that nation states establish for managing migration. The principle that migration should be productive and orderly, for example, is regularly used to justify forms of migrant selection and exclusion that include assessments of 'human capital', such as education, skills and experience, as well the biopolitical qualities of individual migrants that are ascribed to race, gender and age as well as other axes of social difference. The effects are obvious in highly gendered and racialised patterns of migration (Silvey 2004) but also in the ways in which the conditions of migration – creating categories of entry, defining work relations and limiting social interaction – influence how migrant lives unfold (Anderson 2010). Migration policies can hence simultaneously encompass a 'war for talent' where special privileges are used to attract and retain the most 'desirable' and purportedly productive subjects, while labour, marriage, asylum and other kinds of migration that are deemed necessary but as less desirable are exposed to varying forms of differential exclusion both in migration and in daily life (Ye 2016a; Yeoh 2006).

In this chapter I set out to problematise these political rationalities of migration management and their theoretical underpinnings within mainstream migration studies. Both the policy orthodoxy of migration management and mainstream migration scholarship have tended to reduce migration to economic logics that render it amenable to management and the diminishment of migrant agency. In contrast, the approach developed here embraces the excesses and uncertainties involved in migrant mobilities, undisciplined movement across borders and the role of migration in constituting the places migrants move through (a task now also increasingly taken up in critical migration and border studies, see Casas-Cortes et al. 2015). I focus in particular on developing a theoretical foundation based on desire, assemblage and encounter as vocabulary for re-examining migration and its relationship to urbanisation.

I begin this chapter by first introducing emergent scholarship on migration regimes as a means to critically examine contemporary migration patterns in Asia including South Korea. This approach draws attention to the multiple actors involved in shaping migration patterns and regulations. It also looks at the ways in which gender, race, nationality and other axes of social differences are deployed and extended in the stratification of migration and the embeddedness of migration in particular imaginations of the place of migration in national and transnational spaces. While a migration regimes approach offers a useful response to the programmatic claims about the benefits of migration management it also remains fixed within a focus on hegemonic power relations and lacks the spatial depth needed to address questions of agentive will in migrant lives. Accordingly, in the second half of the chapter I develop an alternative set of conceptual tools through the vocabulary of desire, assemblage and encounter. Drawing on both extant scholarship on migration, cities and diversity and the philosophical claims of Deleuze and Guattari (1983, 1986, 1987), I suggest that this focus on desire, assemblage and encounter offers scope to reassess the generation of migration beyond economic rationality, the ways in which this is shaped by shifting urban forms and migration regimes and has consequent impacts on the everyday lives and encounters of migrants in the city.

2.1   Migration Regimes and the Stratification of Movement in Asia

There has been a growing focus in critical migration studies in the twenty-first century on the role of nation states in regulating and shaping migration and a renewed emphasis on borders and enforcement (Bauder 2016). These developments reflect both empirical observations of increasing regulation of migration, such as the propagation of migration management, as well as recognition by scholars that earlier claims about the normalisation and accessibility of migration and mobility have been overstated if not false (Pécoud 2014). While it is now cheaper and faster to move within and between nations than ever before, and communication possibilities between places have proliferated, it is also clear that these opportunities for mobility are not shared evenly between populations or across geographies. Rather, increased speed and efficiency of movement is accompanied by complete immobilisation in the restriction of movement, and much transnational mobility is channelled through highly regulated means of migration and the unauthorised or undocumented forms that necessarily accompany such controls.

In this context, migration scholars have begun to focus increasingly on the establishment of 'regimes' involved in shaping diverse mobilities (Glick Schiller & Salazar 2013), migration management (Geiger & Pécoud 2013) and border control (Mezzadra & Neilson 2013). The focus on such regimes directs our attention to the modes of rule or government that are involved in making possible, directing and setting limits on migration. It involves drawing attention to the work of individual states as well as changing international regulatory systems and their effects in terms of individual mobility in the world. At the same time, however, a focus on regimes exceeds a narrower emphasis only on nation state and individual actors by incorporating analysis of different levels of governance, other factors involved in shaping migration including imagination and representations, as well as social relations, norms and values (Ho, E. 2014). Regimes, then, are recognised as an ensemble of practices that may be characterised by more or less order, constructed through responses to emergent challenges and filled with gaps and potential problems (Casas-Cortes et al. 2015).

Glick Schiller and Salazar (2013), who have offered a sustained development of 'regimes of mobility', highlight several key features of an emphasis on regimes. First, they note that a focus on regimes makes it possible to qualify understanding of the normalisation of mobility with recognition that forms of movement are necessarily stratified. Mobility and indeed migration is situated within social fields of differentiated power where 'disparities, inequalities, racialized representations and national mythscapes' serve to 'facilitate and legitimate differential mobility and fixity' (Glick Schiller & Salazar 2013: 183). Second, they suggest that this means that rather than a singular mobility or migration regime, scholars need to attend to multiple different intersecting regimes that emerge in relation to governing different populations or cut across traditional territorial borders between nations. Such regimes are constituted principally by nation states but also involve a multitude of other actors, including others involved directly in migratory processes such as migrants, families, communities, intermediaries of various kinds as well more indirect players such as international organisations, educational institutions and producers and distributors of cultural forms. Last, they note the importance of imaginaries of migration, not only in relation to relatively fixed conceptions of nations but also in terms of the desirability of different forms of migration and of particular migrants as desirable subjects.

The growing significance of migration in the Asian region offers a useful set of examples through which to develop these ideas around migration regimes and their role in shaping the form and experience of mobility. Over the last two decades, the number of international migrants in Asia has increased by over 40% from 50 million in 2000 to 71 million in 2013, making it the second largest host region after Europe (UN 2015). This reflects trends established in the 1970s as the political–economic and demographic trajectories of countries in the region began to diverge, particularly in terms of the availability and demand for labour (Hugo 2006). Manifest initially in the movement of workers from all regions of Asia into the oil-rich nations of the Middle East, inward migration has since the 1980s emerged as a significant feature of newly industrialised countries in East and South East Asia. By the end of the 1980s well over two million migrants were leaving their countries of origin annually, a figure that has increased to some four million in the three decades since (Asis & Piper 2008; Battistella 2014).

Migration in Asia is largely intra-regional and is characterised by marked geographical contours (Battistella 2014; Collins, Lai & Yeoh 2013). Large numbers of migrants originate from particular countries such as Indonesia, Philippines and Vietnam in South East Asia, India, Pakistan and Bangladesh in South Asia and China. By contrast, a small number of economically advanced nations receive most migrants – Hong Kong, Japan, Singapore, South Korea and Taiwan – with countries like Malaysia and Thailand both sending and receiving migrants. In addition to reflecting the geo-economic make-up of the region, these patterns also highlight important connections between international and internal migration (Amrith 2011). Over the course of the mid-twentieth century, most migration in Asia was in fact internal to newly emerging nation states and was caught up in the modernising and urbanising developmental projects of postcolonial nation building (Asis & Piper 2008). Rural–urban migration has been a significant driver of metropolitan growth and the emergence of megacities in Asia. Urban settings are also important sites for generating outward migration, as rural migrants develop skills and aspirations for further mobility and opportunity across borders. In global cities like Seoul, Singapore, Taipei and Tokyo, the contours of urbanisation generate outward migration but they have also created new opportunities for overseas migrants as labour markets have become increasingly bifurcated (Lim 2003; Yeoh 2006).

These characteristics speak to the ways in which migration emerged in relation to the historical development of nation states in Asia (Yeoh 2017). Indeed, when labour migration flows started to become established in the 1970s and 1980s many countries in Asia were still in the midst of nation-building projects, following postcolonial independence in Southeast Asia or the consolidation of ethnically-scripted and newly democratic nationhood in East Asian polities such as South Korea and Taiwan (Collins, Lai & Yeoh 2013). As destinations, these countries have maintained a strong emphasis on temporary systems of regulating labour migration from other parts of Asia in a way that reflects the demands of nation building, restricting migrants to a transient status where their presence does not further complicate the imaginary of the nation state in the making. Accordingly, the regimes for instigating and regulating migration in Asia by and large preclude 'settlement, family reunification and long-term integration, including acquisition of citizenship, for less skilled migrants' (Asis & Battistella, 2013: 32). It is only in regards to more recent patterns of high-skilled and educational migration, where migration can be aligned with imaginaries of achieving global status for cities or nations, that alternative modes of incorporation have become more apparent (Yeoh & Huang 2011).

A key component of migration regimes in Asia, then, is the establishment of 'hierarchies of regulation … that treat migrants differently depending on ethnicity or level of education and skills' (Lindquist, Xiang & Yeoh 2012: 11). Migrants are not governed by one but rather multiple intersecting regimes where criteria and conditions of entry vary considerably according to the perceived desirability and necessity of different migrants. Gender, nationality, class, race, age and other axes of social difference intersect here in shaping migration and influencing the place of migrants in everyday life. Scholars working on migration to Singapore (Kitiarsa 2008; Platt et al. 2017; Ye 2014, 2016a; Yeoh 2004, 2006; Yeoh & Huang 1998), for example, have demonstrated the ways in which the city-state's increasing reliance on migrant labour has been articulated through intricate examinations and determinations of the biopolitical value of different migrants. Here foreign domestic workers, exclusively female and predominantly Filipino and Indonesian, experience life in the city in a way that is tied to the space of the household and the diminishment of care-work, constraining many although not all opportunities for expressions in public space (Yeoh & Huang 1998). By contrast, in the migration of male construction workers, masculinity intersects with migrant categories rendered through conceptions of race, class and nationality that position (labouring) male migrants as distinct subjects in need of management in public space (Ye 2016a; Kitiarsa 2008). As Ye (2014: 1024) argues, then, 'migration does not simply unsettle or reproduce local gender regimes but it is [rather] a vital part of the social and cultural processes of subjectivation'. Interlaced with the classed character of migrant recruitment, gender, ethnicity, nationality and other social differences contribute to a complex and intersectional division of labour and life where migrant subjectivities are literally 'multiplied' in ways that establish their desirability and appropriate place in society (Mezzadra & Neilson 2013).

These migration regimes are part of a striking formalisation of migration in Asia over the last decade that involve a wide range of actors. Nation states have been pivotal in this process and in particular in establishing markets for temporary and circular migration that are closely regulated and subject to ongoing governmental monitoring (Battistella 2014). In many cases, formalisation has also involved a broader set of bilateral and multilateral agreements in which sending and receiving states and international organisations have coordinated their efforts to shape the outcomes of migration. A focus on circular migration has been particularly prominent, an approach that is promoted as a win-win-win scenario for sending and receiving states but also migrants themselves. South Korea's own EPS provides an indicative example (see Chapter 3 for further details). Established as a means to protect migrant rights, enhance labour market efficiency and reduce unauthorised migration, the EPS operates through government-to-government agreements with 15 sending nations in Asia who manage the recruitment, training and deployment of potential migrants. In South Korea, the EPS stipulates quotas for the nationalities of migrants and the industries and employers they can work for, provides workplace training and protection and secures the departure of migrants at the end of their time-limited visas (KoILaF 2007). Such regimes then also incorporate the interests of sending nation states to regulate migration as part of an export strategy for national economic development. In the process, migrants are transformed into temporary guest workers and are effectively subjected to a form of regulated return (Xiang 2012).

Private intermediation plays an important role in these processes of formalisation, even in contexts like South Korea where the EPS has been designed to exclude migrant brokers. Indeed, one of the features of increasing centralisation of state control in Asian migration is the concomitant importance of a much wider infrastructure of mobility. Through this infrastructure state practices are articulated with institutions, actors and networks that play roles in moving migrants across borders and setting the conditions for their life abroad (Xiang & Lindquist 2014). Profit-seeking migration brokers that sit outside of the state have been particularly prominent subjects of recent scholarship (de Dios 2016; Lindquist, Xiang & Yeoh 2012; Xiang 2012) but it is critical to recognise the wide varieties of intermediaries involved in generating and organising mobility. The EPS, for example, involves a diverse network of actors and institutions involved in various stages of the migration process: from language training centres and insurance providers in sending countries, private companies contracted by the Korean government to train workers, to NGOs contracted to oversee worker welfare (Seo & Skelton 2016). Outsourcing of this kind is commonplace across Asian migrations as governments seek to formalise and integrate different features of the migration process, monitor and manage the position of migrants in society and reduce irregularity.

The focus of much of the growing formalisation of migration and in particular the rigidity of regulation has been migrants who are classed as 'low-skilled' and as a result often deemed as in need of protection and management (Kim 2012). More middling groups, like the English teachers and even international students discussed in this book, do not fit neatly within this process but neither are they necessarily treated in the same manner as purported 'foreign talent'. The migration of English teachers, for example, is also governed by a specific migration regime that ties their presence to a particular labour market niche and sets specific conditions for entry and exit (Collins 2014b; Lan 2011). Notions of nationality, race, age and gender are manifest in this process, particularly in the idealisation of certain kinds of subjects – especially white North Americans – as ideal teachers (Ahn 2014). International students likewise are obliged to be enrolled in particular types of courses and to limit involvement in other activities and, depending on the conditions of their study, may be encouraged to remain or leave following graduation (Shin & Choi 2015). Intermediaries of various kinds are active in these spaces too, configuring the directions and outcomes of migration although often in ways that place the financial burden on employers and educational institutions rather than migrants. As Chapters 5 and 6 demonstrate, both groups experience considerable freedom in their migration and daily lives, especially in comparison to migrants moving through the EPS regime, but they too are subject to the imperative towards management and their narratives reveal both frictions and precarity in movement and everyday life in the city.

As this schematic account demonstrates, the focus on migration regimes offers substantial advancements to understanding the politics of migration and the ways in which it is infused with power relations. This approach serves as a particularly important corrective to the programmatic accounts of migration processes and responses in discourses of migration management. By highlighting the active arrangement of migration, a migration regimes approach rejects the notion that there is simply extant 'emigration pressure' in need of regulation. Rather it highlights the fundamental role of regimes and their multiple actors and organisations in establishing and sustaining particular migration patterns that are characterised by high levels of inequality.

The tendency embedded in this approach, however, is to focus on the structuring of migration and the manner in which it is established within hegemonic arrangements. As Glick Schiller and Salazar (2013: 189, emphasis added) note, for example, the 'term "regime" calls attention to the role both of individual states and of changing international regulatory and surveillance administrations that *affect* individual mobility'. Unlike a focus on migration systems, or orthodox economic renderings of migration, there is greater suppleness in terms of examining the range of actors involved in migration and the varied ways in which movement is enabled and governed. Yet, the emphasis remains on governance itself without adequately addressing the role of migrants, or more importantly the ways in which people on the move not only subscribe and submit to but also disrupt and exceed what is expected of them in migration regimes. There are not the tools here to unpack the ways in which subjectivity is articulated in and transformed through migration, the way migrant lives are 'constituted through a range of intersecting, sometimes competing, forces and processes, and as playing agentic roles in these processes' (Silvey 2004: 499). A migration regimes approach also has limited spatial depth in that its analytical imperatives appear tied to an albeit critical and multi-scalar analysis of nation states and migration. What's less evident is how imaginings of migration circulate beyond apparent regimes, how migration is embedded in social lives and networks far distant from the interventions of sending or receiving states, or the way in which migration articulates not just in relation to borders but across the diverse terrains of migrant mobility in villages, towns and cities and beyond. It is, in other words, unclear to what extent a migration regimes approach permits the figure of the migrant to have 'its own history and social force' (Nail 2015: 5). To address subjectivity and its articulation across the variegated spatialities of migration it is necessary to recentre migrant lives and to consider the ways in which migration emerges in specific configurations of place and possibility. To work towards this goal the following three sections develop an approach centred on the vocabulary of desire, assemblage and encounter and their possibilities for rethinking migration.

2.2   Desiring Migration

The migration regimes that have been established in South Korea and the Asian region over the last three decades play a significant role in constituting a politics of migration that is animated by aspiration and anxiety and represented along lines of gendered, racial and other hierarchies. Migration expresses aspirations by nation states to be part of flows of knowledge, capital and actors and a wider national desire for economic development and global emergence. Within such aspirations migration is promoted but only in relation to circumscribed identity positions where migrants are framed differentially as diasporic returnees, mobile professionals, migrant labour, marriage migrants, international students or other migratory figures. A migration regimes approach tells us much about how such arrangements come to be formed and how they play a role in establishing graduated forms of incorporation where migrants are marked as more or less desirable forms of labour, as amplified symbols of global status or derided and made invisible through their marginalisation (Ong 2007). What is far less apparent in this focus on migration regimes is the manner in which people on the move position themselves in relation to migration and in the process become *desiring* as well as *(un)desirable* subjects. Without addressing the question of what instigates and sustains migration there is a risk that scholars implicitly accept or even reproduce utility-maximisation models (De Haas 2011) that see migrants as transnational careerists who are only subject to the forces of economics and the direction of nation states (Wang 2013).

Following the insights developed initially by feminist scholars of migration (Silvey 2004) and arguments for renewed emphasis on emotions and affect (Conradson & McKay 2007) it is possible to develop a focus on migrant subjectivity and the spatialities of migration that 'moves beyond utilitarian questions of risks and rewards, costs and benefits, to a consideration of the overflow of [migrant] aspirations' (Chu 2010: 5). This involves grappling with the ways in which rationales can emerge through feelings or embodied responses to ideas about migration that emerge in social networks and representations (Kim, Y. 2013; Salazar 2011). Migration emerges in relation to local, national and transnational flows of images through media (Mills 1999; Sun 2002), social networks that connect people across borders (Hoang 2015; Chu 2010) or the workings of nation states and others involved in the promotion of migration (Cheng 2011; Choy 2003; Yang 2016). The engagement with such imaginings and their articulation with the actualised experiences of migration also involve refigurations of subjectivity, not as a self-conscious choice about identity positions but rather in the sense that migration involves unpredictable encounters that literally alter our own understandings of the world. Transformations of subjectivity of this kind are of course often gendered and racialised, not least through the ways in which the position of migrants is scripted in relation to shifting notions of femininity (Kim 2013; Mills 1999) and masculinity (Ahmad 2009; Kitiarsa 2008). There is agency in migration, then, but scholars need to consider the ways in which this is socially distributed and unevenly enabled through the wide range of relationships and encounters that characterise processes of migration. As Conradson and McKay (2007: 172) put it, 'feelings may act to enhance and secure particular forms of transnational labour mobility, whilst also disrupting or undermining capitalist economic logic'. Put otherwise, migrants are both subject to and also rework the territorialising powers of migration regimes – they become both the labouring and learning bodies desired in these regimes but also active and desiring human subjects whose presence can never be completely contained.

A focus on desire as a conceptual vocabulary for migration studies, or more precisely on the ways in which migration can be an expression of desire, provides a productive avenue for elaborating these claims (Collins 2017). My own approach to developing desire as a conceptual vocabulary draws on the philosophy of Deleuze and Guattari (1983) but it is also situated in relation to several other recent usages of the term desire in accounts of migration in the Asian region. The notion of desire has been particularly productive for scholars exploring the growing mobility of women in Asia and the ways in which such movements are tied up in discourses and possibilities for becoming modern and independent. Mills' (1999) account of Thai women moving from villages to Bangkok is a particularly powerful example. Her ethnography reveals how consumption, independence, family obligation and a desire to become modern coalesce in accounts of mobility to the city as well as the potential of these undertakings to 'confront and potentially transform existing social discourses about appropriate roles and (gendered) identities' (Mills 1999: 183). Desire can be transformative, then, as Kim's (2013) account of women from Japan, China and South Korea pursuing western higher education also suggests. However, while these educational migrations reveal desires to 'have a degree of control over the conditions of their individual lives' her research also argues that the discourses that support and encourage such undertakings emerge in 'flows of desire that now operate transnationally' (Kim 2013: 2). Our desires then are not necessarily our own (Berlant 2011) and need to be approached in a situated manner or what Chu (2010: 5, emphasis in original) describes as 'the *pragmatics* of desire – the cultural–historical configuration of its incitements, the social hazards of its translation into action, the political implications of its effects'.

In the South Korean context, Cheng's (2011) account of migrant entertainers and the US military offers a demonstration of some of the complex ways in which desire comes to manifest in relation to migration. Her research traces the lives of Filipina migrant women who work as entertainers at clubs in *gijichon*, US military camps in South Korea. In contrast to framings of these women only as victims of sex trafficking, simultaneously desired and undesirable, Cheng (2011) draws on notions of desire, as well as wider engagements with agency in migration, to incorporate not only the racialised and gendered landscape of migrant work but also the manner in which migration involves aspirations for alternative futures and meaningful relationships. Here a focus on desire serves as an antidote to a reading of migrants as victims because it demonstrates that even as these women are framed as objects of desire they are also subjects of desire:

> The women bring into their workplaces and their migrations their own romantic and erotic imaginings, continuously blurring the boundary not only between labor and love, but also between the public and private, and between performance and authenticity' (Cheng 2011: 10)

This account then demonstrates how the desires of migrants can exceed or subvert the expectations of migration regimes. Like other scholars of migration drawing on notions of desire (Chu 2010; Hoang 2015; Kim, Y. 2013; Mills 1999), Cheng (2011: 75) also recognises the key role played by different state actors and wider imaginative geographies 'in the cultural production of this subjectivity of desire' that facilitates and sustains migration. Desire emerges as a contested field of opportunity here, then, where straightforward narratives of choice and coercion do not provide adequate explanations for the instigation of migration, nor the way it is managed and responded to by states, migrants, employers, NGOs and other actors.

One of the key insights that emerge from these diverse accounts of migration and gender in Asia is that a focus on desire needs to do more than just speak of varying objects of desire, the different things that individuals, migrants or not, may *want*. Rather, drawing attention to desire in relation to migration demands attention on *desiring*, the manner in which we develop attachments to particular things, imaginings, people and places as desirable (Berlant 2011). Recognising this means that desire itself may be incoherent, free-flowing or shape-shifting while also being a powerful driving force in migration and other agentive undertakings. The examples raised by scholars – of becoming modern and urbane (Mills 1999), negotiating sexual subjectivity (Cheng 2011), integrating educational, cultural and individualising desires (Kim 2013) – reveal some of this where desiring intersects with uncertainty, anxiety and other complex feelings about the future implications of present actions.

This is also a theme that runs through the reading of desire as a social force developed by Gilles Deleuze and Felix Guattari in *Anti-Oedipus* (1983), *Kafka* (1986) and to a lesser degree *A Thousand Plateaus* (1987). Deleuze and Guattari (1983) centre much of their critique of psychoanalysis on a reading of the centrality of desire, going so far as to suggest that 'there is only desire and the social, and nothing else'. By this of course they mean to argue that desire is a social force that animates life and undergirds actions that people undertake and the relationships they establish with places, people, ideas and objects. Their account is positioned specifically in opposition to psychoanalytic readings of desire that begin with Freud and Lacan, who read desire as something that is an effect of power relations and something that will inevitably be characterised by lack, by a frustration that emerges when we can't get what we want, the objects of our desire (Butler 1987). The notion that desire is a social force that animates the world and our actions within it, *desiring*, departs from this psychoanalytic view because it highlights how desires are not our own – we do not control objects of desire – but rather they are part of wider representational and embodied schema and must accordingly be understood in terms of particular social formations. In migration, for example, desire for expressions of improved livelihood or freedom, as demonstrated in the accounts above, may serve as the impetus for movement but in this process desire also draws migrants into connection with other places, altering their socio-cultural arrangements and generating new encounters between migrants and locals.

This emphasis on desire has considerable potential to advance understandings of migration in a way that addresses the multiscalarity and temporality of mobility and its articulation across migrant bodies and social relations, through places of work and study, to urban, national, regional and global spaces. First, focusing on desire makes it possible to displace the myth that migration follows only from rational choices that are made at particular points in time (De Haas 2011). While it is certainly feasible to assert that a person has an 'interest' in migration and seeks to realise this through particular actions, for Deleuze and Guattari (1983, 1986) such interest needs to be conceived as part of the social contexts that individuals inhabit and becomes possible because of desires people invest in the value of that social formation (Smith 2007). Migration, including the forms of labour and educational migration that are discussed in later chapters, is often coded in the norms of contemporary capitalism and the reification of economic, social and cultural power that is reproduced through this assemblage. Accordingly, the notion of economic success and livelihood improvements produced through migration need to be situated analytically in relation to the representational regimes that support and shape how we know that migration makes optimistic futures possible. At the same time, migration can also be generated through other desires that emerge from dissimilar if not fully distinct assemblages – desires to become modern, feminine or masculine, to embody identities affiliated with being a 'global subject' or 'cosmopolitan', obligations to community, place or feelings of filial piety, or a will to avoid, subvert or escape from social restrictions and institutional constraints (Kim, S.K. 2013; Phan 2016). Migration in this sense is necessarily generated through multiple rather than singular forces and in an ongoing process where past experiences and memories, present actions and circumstances and future possibilities are interlinked in the emergence of migrant lives (McCormack & Schwanen 2011). Becoming a migrant involves a complex interplay of these forces and their coalescence around strategic planning and opportunism that take shape in efforts to achieve or avoid certain kinds of futures.

Second, focusing on desire in migration highlights the way in which the drivers of migration are constructed, assembled and arranged. The drive or stimulus for migration does not originate from a particular object (a predetermined income or amount of money, an overseas qualification) or subject (a successful migrant returnee) but rather emerges through the ways in which these things are discursively produced (Bignall & Patton 2010). Desire, as Berlant (2011: 16) puts it, is 'a cluster of promises'. In this respect, desire can be understood as influenced by discourses that establish certain actions, undertakings, objects, behaviours or opportunities as desirable while coding others as less desirable. As discussed (Glick Schiller & Salazar 2013), these imaginaries are constructed by a wide range of actors involved in the formation of migration regimes including nation states seeking to activate migration to address labour shortages or demographic decline (Lukacs 2015), to attract and capture foreign 'talent' or 'experts' (Ong 2007) or to capitalise on multicultural presence as part of designs for national emergence (Ahn 2012; Doty 2003). In addition to their role in arranging and stratifying migration flows, then, the migration regimes discussed also contribute to enlisting migrants in the promise of migration. This can occur directly, through practices of state marketing like those that can be found on the websites for the EPS, the official English Program in Korea, or the Study Korea Project. Alternatively, migration can be produced through a broader social assemblage of actors including migration agents and other intermediaries (Lindquist, Xiang & Yeoh 2012), the cultural productions of South Korea in TV, film and music (Shim 2006), social networks of family and friends (McKay 2007), promotion in sending states (Collins et al. 2014) and wider cultural politics that valorise mobility and migration as developmental trajectories (Tyner 2013). None of this suggests that migrant desires simply equate to these discourses or that their effect is to 'recruit' migrants for the objectives of the nation state, employers, universities or intermediaries. Rather the focus on the arrangement of the drivers of migration emphasises that desire forms through inputs from a range of ideas and actors that can be both complimentary as well as contradictory in the ways in which they shape the form and processes of movement.

Last, drawing attention to desire places emphasis on the ways in which migration can be transformative, the practical as well as political dimensions of migration for people, communities and nations. Deleuze and Guattari (1983, 1986) insist on this when they suggest that 'the analysis of desire, is immediately practical and political', it is 'the art of the new', it 'actively participates in the drawing of the lines' between bodies, of creating and destroying assemblages. Of course the idea that migration is a transformative process is not new; migration involves change for people on the move in terms of the places they live in the world, ideas about themselves and possibilities for the future (Castles 2010). It has been common, however, to view transformations as something that occurs after and as a result of migration, rather than seeing the emergence of migration as transformative in and of itself. Likewise, where migration is understood as transformative for nation states it relates to subsequent effects in society – the diversity of populations, open or xenophobic responses, changing labour markets or new ethnic spaces in cities – rather than seeing migration and its consequences as part of interlinked processes from the outset. By contrast, the emphasis on desire employed here posits migration as a process of desiring production that is itself an expression of transformation, evidence of deterritorialising and reterritorialising processes for both people on the move and the places they traverse and inhabit through migration. As scholars working on migration in other parts of Asia have demonstrated, the transformative potential of migration is not a certain and predictable outcome, rather it is also caught up in anxiety and uncertainty (Lindquist 2009) and negotiations of gendered, classed and racialized subjectivities (Cheng 2011). Moreover, despite the effort migration regimes invest in enlisting migrants and shaping pathways – arrival, work, study, departure – there is substantial variation in manifestations of mobility as migrants become irregular, subvert rules, transition between visas, depart early, demand rights or avoid support (Bauder 2016; Papadopoulos, Stephenson & Tsianos 2008; Robertson 2013). There is, then, considerable unpredictability in migration – it creates new encounters between mobile and local subjects, it raises questions about citizenship and politics, and alters the expectations and subjective positions that migrants inhabit.

There is unpredictability that emerges in migratory processes and the way in which desires to circulate are entwined in transformations of migrant selves, in the social spaces that migrants inhabit and the places where migration has impacts such as the city of Seoul. The desires to circulate that emerge through the narratives of migrant workers, English teachers and international students discussed in this book speak to the possibilities that are invested in migration itself but they also suggest a more open and emergent conception of desiring to become in circulation. Taking this focus on desire to discussions of migration helps to transcend the emphasis on rational choice and the demand for labour that remains a substantive feature of migration scholarship, not least in work on labour migrants in South Korea (Kim, W.B. 2004; Jun, Ha & Jeong 2013; Song 2015). It also draws attention to the role of the state, and wider practices of statecraft in assembling the possibilities for migration and for both allowing the flow of human beings across borders as well as seeking to shape and circumscribe their presence. As Sections 2.3 and 2.4 demonstrate, a focus on these different articulations of desire also open up questions about the ways in which the traversal characteristics of migration can be part of the composition of urban, national and transnational spaces and the politics of encounter that characterise the flows of everyday life in the city.

2.3   Urban, National and Transnational Assemblages

One of the features that both programmatic accounts of migration management (e.g. Ghosh 2007) and many critical perspectives on migration regimes (e.g. Glick Schiller & Salazar 2013) share is a relatively limited conception of the spatialities of migration. For advocates of migration management this is obvious in the pre-eminent status given to states as the sole authority controlling national territories that rest within broader stable international relationships – a nested geography reminiscent of what Agnew (2003) described as 'the modern geopolitical imagination'. The accounts of migration regimes described in Section 2.1 have much more nuance to their reading of nation states, not least in recognising the non-monolithic character of the state, the multiple actors involved in migration and the uneven stretch or reach of different migration regimes. Yet, as I have already noted, a migration regimes approach still turns our attention to the state first and foremost and governance more widely, observing mutations and manoeuvring in relation to migration but not necessarily revealing the diversity and depth of spatial relationships that are involved (Xiang & Lindquist 2014).

The focus on desire and migration, or more precisely *desiring* migration, elaborated in Section 2.2 goes some way to opening up possibilities for a more multidimensional spatiality of migration (Dovey 2011). This is because, as the examples in Section 2.2 demonstrate (Cheng 2011; Kim 2011; Mills 1999), a focus on desire reveals the integration of sociality and spatiality in migration and other phenomena. Kim's (2011) account reveals the transnational imaginative geographies of popular culture and media that generate and sustain impulses for mobility; Mills (1999) shows how migration is embedded in the promise of alternative subjectivities that are materialised in specific places like Bangkok; and Cheng (2011) helps us to see how the micro-politics of arranging and negotiating spaces such as the military camp are interlinked into processes of becoming that cannot be disentangled from the territorialisation of the nation state. While not necessarily foregrounding space or spatiality in their own vocabularies the accounts presented by these authors provide substantial evidence towards an argument that migration cannot be understood as simply movement between distinct nation states (c.f. migration management) or that migration is always and primarily oriented towards its governance (c.f. migration regimes).

The approach taken in this book is to focus on the city and urban space as geographies through which migration, desire and politics are manifested. The point of this focus on the urban is not to diminish the importance of other spatial formations in migration, whether the less urban emergence of migrant lives and aspirations (Shubin 2012), or the national and transnational arrangements involved in assembling mobility (Collyer & King 2015; Xiang & Lindquist 2014). Rather, my focus on the urban reflects the empirical significance of Seoul in the patterning and experiences of migration in South Korea over the last two decades, and the wider importance of major cities in Asia as sites through which migration has been arranged and where the politics of migration emerges in its most immediate forms (Wong & Rigg 2010). In such cities, where large foreign-born populations have emerged in recent years, it is crucial to attend to the manner in which migratory flows and relations are territorialised in the fabric of urban life through relations that extend well beyond the jurisdictional limits or lived experiences of the city.

Assemblage thinking, which has become a prominent feature of recent urban scholarship (McFarlane 2011a, 2011b; Edensor 2011; Farias & Bender 2012; McCann & Ward 2012), provides a useful conceptual vocabulary for interlinking analysis of the city into wider socio-spatial arrangements. Resisting orthodox conceptualisations of 'the city' as a bounded unit and stable object, assemblage thinking draws attention to the relational constitution of cities, the question of how urban fabric is put together, how it is actively assembled and disassembled. Assemblage thinking captures processes of composition, the ways in which things are 'gathered together, aligned, transformed or dispersed'

(McFarlane 2011a: 667) and critically the way in which different elements interact, the capacities and forces that are generated therein. Deleuze and Guattari (1987), whose later work serves as the template for much geographical articulation of assemblage (Anderson & McFarlane 2011), emphasize that assemblages – nations, cities, communities, bodies, machines – need to be understood not as the sum of their parts but rather as the result of particular types of co-functioning:

> What is an assemblage? It is a multiplicity which is made up of heterogeneous terms and which establishes liaisons, relations between them, across ages, sexes and reigns – different natures. Thus the assemblage's only unity is that of a co-functioning (Deleuze & Parnet 2007: 69).

The advantage of such an approach is that it does not essentialise the city as a pre-existing entity, nor privilege particular perspectives as the key means of examining how cities or other socio-spatial arrangements come into being, are experienced or reconstructed (Simone 2011). For my purposes the language of assemblage also offers an entry point to address national and transnational socio-spatial formations that matter immensely in the study of migration and in the ways in which migrant lives contribute to the urban fabric of cities like Seoul. First, speaking of urban or other assemblages draws our attention to the manner in which the city can reach out through imaginative and material connections that build relations with other people and places across vast geographical distances. In migration, for example, people are drawn into relations with the possibilities of places – the representational promise that expresses desire for migration. Cities are a significant part of this desire because the imagination of urban life is something that is often articulated through ideas of becoming modern and achieving social and economic advancement (Mills 1999; Shi & Collins 2017). The image of the city, and its anticipated or experienced material manifestations, then, articulate with transnational socio-spatial formations.

Seoul's transnational urbanism manifests strikingly in the way in which the cultural, economic and material products of this city have become embedded in processes of production and consumption across Asia that have significant importance in processes of migration. Hindman and Oppenheim (2014) demonstrate an excellent example of these growing connections as they trace the paths of consumer goods from Seoul's markets to shops in Kathmandu and the associated rise of notions of 'Korean quality'. Rather than a unidirectional movement of goods and associated symbolic value, however, Hindman and Oppenheim (2014) illustrate that contemporary consumer desires for Korean goods are also entangled in conceptions of urban modernity that echo through patterns of labour migration from Nepal to factories on the outskirts of Seoul's metropolitan region (see also Seo & Skelton 2016). Likewise, for many of the young people from Asia whose stories are foregrounded in the chapters later in this book, migration to

Seoul and South Korea, whether to work or study is bound up in narratives of South Korea's seemingly exceptional development trajectory, the lessons it offers for South Korea's neighbours and the prospects for becoming through migration. While such comparative frames manifest differently for many English teachers, narratives of South Korea or Asia more generally as a place of orientalist desire pronounced in their stories also speak to the ways that cultural and material iterations of place generate migration through their extension across territories. These transnational circulations of South Korea and Seoul conjoin 'contemporary consumerism with shifts and displacements engendered not only by transnational media and its reception, but also by regimes regulating transnational labor migration' that articulate with shifting political economies of migrants' points of departure (Hindman & Oppenheim 2014: 468).

While the circulation of Seoul's imaginative and material potential speaks to a transnational dimension to urban life, assemblage thinking also draws our attention to processes of territorialisation, the stabilisation of different socio-spatial arrangements. Cities such as Seoul may extend in imaginative and material ways transnationally but the movement of people to and from Seoul is also regulated in ways that seek to determine what kinds of migration are acceptable, channel the movement of migrants and set limitations on the place of migrants in urban life and society more generally. Urban and transnational arrangements, then, necessarily link with national regimes for governing migration. Mezzadra and Neilson (2013) capture an important element of this when they discuss the internalisation of border practices. They argue that it is necessary to examine the ways in which national borders reach internally within nations and shape the flows of everyday life: 'establishing internal administrative boundaries and categories that divide labor markets, separate migrant groups beyond and within the boundaries of ethnicity, and provide parameters within which individual migrants negotiate their biographies' (Mezzadra & Neilson 2013: 138).

The significance of national regimes for governing migration in the shaping of urban life is apparent in a substantial literature in Asian cities on the experiences of migrant workers and their participation in place making practices (Law 2002; Wong & Rigg 2010; Yeoh & Huang 1998; Ye 2016a). Here scholars have explored how the socio-legal status of different migrants influences their capacity to navigate urban spaces and to establish meaningful social relations. Yeoh and Huang (1998) provided a striking early example of this in their account of foreign domestic workers in Singapore. Their account highlights how notions of private and public space in the city can be inverted through the socio-legal status of domestic workers: 'her temporary "home" in a foreign land is also her "place of employment" [shaped by] the social relations of waged work';…places away from her employer's home such as public spaces may in fact afford more degrees of "privacy" and personal freedom' (Yeoh & Huang 1998: 585). In a more recent account focused on Nepalese in Seoul, Seo and Skelton (2016) similarly discuss the ways in which the regulations in the EPS migration system serve as mechanisms for controlling the work lives of migrants, and also socially isolating and spatially segregating them in the city.

Like Yeoh and Huang (1998) and other scholars (Elsheshtawy 2008; Law 2002) however, Seo and Skelton (2016) also emphasise the ways in which these workers exercise agency while inhabiting space, especially on their days off work. They describe the vibrancy of an informal Nepal Town where workers gather on the weekend as well as practices of talking, hanging out and loitering that serve as subtle tactics for remaking space and asserting agency that is often denied by migration regimes. Such spaces and practices are transient but they are persistent in that they have become established if time-limited features of urban life in Seoul and other Asian cities (Collins 2012). Accordingly, they demonstrate that 'by "occupying" and "taking over" left-over spaces, low-income migrants [can] circumvent official policy and begin to stake a claim to the city's public space' (Elsheshtawy 2008: 985). In doing so, they provide an indication of the ways in which migrants might then also become involved in remaking urban life in spite of the restrictions imposed by migration regimes, not through official discourses of multiculturalism (Han, G. 2007) or cosmopolitanism (Yeoh 2004) but rather through appropriating urban spaces to their desires.

The relationship between national regimes for governing migration, the transnational movement of migrants and the fabric of everyday life in cities is not stable or straightforward then. They come together through transnational circulations of ideas and people that express forms of desire that interact with and are at times shaped by the national regulation of migration and in turn surface in the everyday life of migrants in specific times and spaces of cities like Seoul. The advantage of assemblage thinking is that it makes it possible to draw these seemingly divergent aspects of migration together and to emphasise the importance of relations and practices, as well as the uneven exercise of power, that matter in the manifestation of different lives and spaces in the city.

2.4   Politics of Encounter

The analytical connection between notions of desire and assemblage rests on carefully examining the *encounter* between individual migrants and urban life, the effects of national migration regimes, and the influence this has on the negotiation of everyday spaces. Encounter is not simply the meeting of two separate entities, migration and cities or people from different places, but rather emphasises a process of becoming and change (McFarlane 2011b; Simone 2010). In migration, as has already been made clear, mobility and encounter with people and places involve potential transformations in subjectivity that cannot be grasped in advance. The encounters involved in migration also hold potential for transformation in the spaces that migrants move through, in the people that they encounter and in the personal and public futures that migration makes possible.

Cities and the micro spaces of urban life are critical sites for the surfacing of such encounters because they are the places that different people meet each other, where migrants encounter others in workplaces, schools and public spaces, and where perceptions and possibilities can be changed. Simone (2010: 191–192) captures this through thinking of cities as made up of different 'crossroads':

> The key is how any place in a city can become a moment and opportunity to create the experience of a crossroads where things intersect – in other words, take the opportunity to change each other around by virtue of being in that space, getting rid of the familiar ways of and plans for doing things and finding new possibilities by virtue of whatever is gathered there.

A focus on the urban lives of migrants then also draws attention to questions about the role of migrants as more than just temporary sojourners but rather as active constituents of urban futures. As the chapters in this book will demonstrate, migration has now become an embedded feature of life in the South Korean capital and the presence of more diverse populations is raising difficult questions about the cultural politics that govern everyday life. 'Non-Korean' residents are increasingly present in the public spaces of the city, from transport, to markets, parks and res-taurants and the cultural sphere of news and popular television (Han, G. 2016). They are also present in 'private' and institutional spaces like workplaces, univer-sities, schools and homes, as workers and students or as spouses and parents of Korean citizens (Freeman 2011). Encounters with difference, then, are also an emergent feature of life in South Korea and particularly in Seoul, raising questions about 'the capacity to live with difference' (Hall 1993: 361) and the potential to incorporate 'foreigners' as ordinary rather than exceptional urban subjects.

The discussion of worker, student and teacher lives in this book draws attention to the cultural politics that emerge in migration and diversity through a focus on encounters in urban life. Racial, gendered or classed differences that are prominent in migration cannot be understood as predetermined in advance only by the dis-courses that frame migrants vis-à-vis locals but rather must be examined through their articulation in encounters with difference that manifest in everyday life. Indeed, it is in the constant encounters that characterise urban life that individual understandings of difference are established, negotiated and in some cases also contested (Young 1990). It may well be that in many instances encounters with others only serve to reinforce prejudice, where pre-existing ideas about different groups are sustained because of the contexts within which encounter occurs (Valentine 2008). Nonetheless, scholars also point to everyday encounters as moments of hope, where the over-determined identities of 'us' and 'them' can be disrupted (Wise 2005), or where strategic attempts can be made to build connec-tions between groups (Amin 2002).

The academic and political lineage of much scholarship around encounter comes from explorations of urban life in Europe, North America and Australasia where conditions and settings of migration and diversity differ significantly from those in South Korea and East Asia more broadly (Collins, Lai & Yeoh 2013; Ye 2016b). Two issues are particularly salient in distinguishing the encounters discussed in this book. First, the accounts of migrants in this book address the ways in which migration is situated within and in many respects rubs up against ongoing efforts to maintain myths of ethnic homogeneity in South Korea (Lie 2014). The narratives of migrants in Chapters 4, 5 and 6 speak to the ways in which notions of ethnic difference and relatively closed social networks demarcate lives and are often reproduced in desires for social connection with co-ethnics. Second, encounters in this research are shaped by the regulatory settings within which migrants arrive and are governed in their day-to-day lives in Seoul. These settings influence where and under what circumstances workers, teachers and students encounter both Koreans and other non-Koreans, whether as employees, tenants, teachers, co-workers, classmates or students. In each of these circumstances status matters and is produced not just through migration categories, but also intersects with individuals' nationality, ethnicity, age, gender and sexuality. As the substantive chapters in this book will show these intersectional differences tend to configure migrants as more or less desirable subjects and to shape the politics of encounter that unfolds in their daily lives.

By focusing on encounters in this book then, I am also concerned with the ways in which these are shaped by the conditions of migration and the material and metaphorical contours of diversity in cities. Encounters take place in what Pratt (1992: 7) and others (Yeoh & Willis 2005) call 'contact-zones', a term that calls attention to 'the spatial and temporal copresence of subjects previously separated by geographic and historical disjunctures'. In such space–times subjects need to be understood through their relations to each other, in terms of 'copresence, interaction, interlocking understandings and practices, often within radically asymmetrical relations of power' (Pratt 1992: 7). The increasing circulation and presence of migrants in the Seoul metropolitan region is generating a multiplicity of contact-zones, either around workplaces and educational institutes or in the public spaces of the city. In all these spaces, encounters are cut through with processes of social differentiation that reflect the privileging of different subjects, their economic, social and cultural resources, and commitments to tolerating or learning with difference (Wilson & Darling 2016; Valentine 2008; Ye 2016b). A focus on encounters draws attention to the influence of both history and present conditions and also serves as a template for highlighting what sorts of circumstances or orientations might support more affirmative or transformative encounters with difference (Wilson & Darling 2016). As the accounts discussed in this book demonstrate, migrant lives are often characterised by difference, distance and dissonance but there are also experiences of generosity, courtesy and cooperation that generate trajectories towards more cosmopolitan futures.

The politics of encounter that are generated in these migratory patterns then need to be understood not simply as temporal disjunctures in the present, but also as constitutive of the temporal horizons for urban futures (Ahmed 2002). The differential mobility and incorporation of workers, teachers and students demonstrates that these groups are not viewed equally, either in terms of their present circumstances or their place in the crafting of Seoul as a global city. We can read this coupling and decoupling of migrant and urban futures through an emphasis on migration regimes, but it is critical to also recognise how these processes are constituted through everyday encounters and what they permit in terms of the establishment of migrants' place in the city, their rights to urban lives and futures (Bauder 2016; Ye 2016a). Moreover, as the narratives discussed here will demonstrate, we cannot read a direct line of causality between the migration regime and migrant presence and practice. Rather, we must also pay attention to the histories of migration and migrant lives, to their convergence in desiring-migration, and their manifestation in urban spaces that are always in the making, always oriented towards necessarily divergent futures. Encounters and the politics that shape them, then, 'exceed the event of the encounter' (Karaman 2012: 1293) because they are not simply interactions in the here and now but rather draw on both spatial and temporal links beyond the city itself. This open conception of migration and urban life informs the analysis undertaken in *Global Asian City* and in particular serves as a template for exploring the convergence of different migrant trajectories and their discrepant experiences of migration and Seoul.

2.5   Conclusion

In this chapter I have developed an argument for a more sophisticated approach to conceptualising the relationship between migration and urban life that makes it possible to move beyond the limitations of both programmatic accounts of migration management and critical perspectives on migration regimes. As I have shown a migration management approach, which is regularly supported by orthodox concepts derived from mainstream migration studies, serves to frame the world of mobility in a way that treats migration as a problem and privileges states and international organisations as arbiters of 'orderly, predictive and productive' forms of mobility. Such an approach clearly simplifies the politics and processes of governing migration and diminishes the agency of migrants negotiating these arrangements and their capacity to transform themselves and the places they move through. The recent focus on migration regimes goes a considerable way to revealing this and to establishing tools for unpacking the multiple actors involved in migration governance and the stretch of national borders and their mechanisms of control. Yet, this approach also retains an emphasis on governance and the state primarily and as a result does not adequately address the agentive capacity of migrants, the multidimensional spatiality of migration or its transformative potential.

The conceptualisation of migration and its linkages to cities developed here offers a different starting point for these questions that centres first on migrant lives, or indeed the lives of those who become migrants and then traces emergent connections to ideas, places and people that manifest in migration. The approach discussed in this chapter makes it possible to show how the generation of migration (desire) occurs across the mobilities and lives of migrants, is shaped by shifting urban, national and transnational socio-spatial forms and regulation (assemblage) and has consequent impacts on the everyday lives and encounters of people in the city, including migrants. One of the features of this approach is that the conceptual vocabulary of desire, assemblage and encounter need to be read in relation to their emergence in empirical accounts and in this book that primarily takes the form of the narratives of different migrants in Seoul; workers, teachers and students. These terms offer a starting point that directs our attention to but does not over-determine our reading of migration and cities, it retains the possibility of addressing the varying and unpredictable inputs and outcomes of migrant lives in Seoul.

In Chapter 3 I begin this process of unpacking the intersecting dimensions of migration and urban life in Seoul through a focus on how contemporary patterns of migration have emerged in relation to shifting migration regimes in South Korea. Extending beyond a reading of only forms of governance itself, however, the chapter explores three different biographies that reveal how shifting imaginative geographies of Seoul and South Korea are generating new forms of migration that come to articulate with these regimes but also lead to considerable divisions in the spatial, social, legal and economic lives of migrants in Seoul. The next three chapters address more directly the ways in which these forms of desiring migration play out in encounters with the city, in the ways in which workers, teachers and students become unevenly part of urban life in Seoul and contribute to shifts in particular urban spaces. Subsequent to this the potential of encounter is addressed more directly in Chapter 7 that questions how actual forms of multicultural presence do and do not align with the assembly of a globally oriented Seoul.

Chapter Three
Migration Regimes, Migrant Biographies and Discrepancy

Let us begin by accepting the notion that although there is an irreducible subjective core to human experience, this experience is also historical and secular, it is accessible to analysis and interpretation, and – centrally important – it is not exhausted by totalizing theories, not marked and limited by doctrinal or national lines, not confined once and for all to analytical constructs.

Edward Said (1993: 35), Culture and Imperialism

As I outlined in Chapter 1, the migration of all kinds of migrants to Seoul over the course of the last two decades are entangled in related transformations of the city and South Korea more generally. Yet, the three groups that are the focus of this book have tended to be treated as distinct and largely incommensurable in their migration, their socio-political status and their experiences. This is apparent across public discourses (Han, G. 2015), policy formulations (Seol 2012) and scholarly accounts (Shin & Choi 2015) where understandings of migration and its relation to the South Korean nation have been established through ideological framings and analytical postures that operate through exclusions. The place and experience of 'migrant workers', 'English teachers' and 'international students' have been treated as 'fundamentally integral, coherent and separate' (Said 1993: 35) and as a result only comprehensible through interpretations that defend the essence of these categorisations. Within these interpretations, 'migrant workers' are unskilled labourers from poor developing countries who are at once victims and criminals (Kim, S. 2012); 'English teachers' are white travelling sojourners from western countries who come and go as they please (Ahn 2014); and

Global Asian City: Migration, Desire and the Politics of Encounter in 21st Century Seoul, First Edition. Francis L. Collins.
© 2018 John Wiley & Sons Ltd. Published 2018 by John Wiley & Sons Ltd.

'international students' are future high-skilled migrants whose inherent human capital is a key force in demographic renewal (Shin & Choi 2015).

Following the line of thinking established by Edward Said in *Culture and Imperialism* my aim is to offer a different reading of these seemingly incommensurable migrations and their place and experience in twenty-first-century Seoul. Said's (1993) 'contrapuntal perspective' is one that advocates a fuller engagement with the entanglements and dependencies of these formations, an analysis that neither essentialises difference nor ignores discrepancy. The migration of these three groups can be viewed as not only interrelated in their chronology, each emerged primarily in the 1990s and early 2000s, but also in their articulation with fundamental transformations in Seoul's position vis-à-vis national, regional and global spaces over the course of the twentieth century. Seoul's emergence as a global centre of commerce, a space for all kinds of transnational flows, has hinged on a different set of relationships with the worlds beyond South Korea's borders and in particular a shifting set of population mobilities. This chapter involves a narrowing of the analytical focus to the specific governmentalities that are involved in the generation and management of migration across these three categorisations and to the biographies of migrants encountering and traversing these regimes. This approach works through an analytical stance of juxtaposition that involves letting different experiences play off each other. The aim is to 'think through and interpret together experiences that are discrepant, each with its particular agenda and pace of development, its own internal formations, its internal coherence and system of external relationships, all of them coexisting and interacting with others' (Said 1993: 36).

In the first part of the chapter, I focus on the shifting contours of migration governmentality in South Korea over the course of the 1990s and the ways in which the state undertook a form of 'strategic ambivalence' (Ybiernas 2013) in its attitudes and responses to all kinds of migration. This ambivalent posture has gradually given way to a much more managerial approach, where the state has targeted different migrations as objects for intervention to address social and economic challenges, and to smooth over political and moral panics. What has emerged is a highly stratified migration regime where migrants are viewed differently by the state depending on their nationality, education and financial status and guided through migration to particular kinds of occupational niches. Recognising the ways in which the migration regime generates discrepancy of this kind the second part of the chapter shifts focus to the biographies of three individuals and their encounter with migration. The focus here is on the imaginative geographies of South Korea and the intermediation involved in making mobility possible. The final part of the chapter draws these biographies and governmentalities into the same frame by accounting for the discrepant mobilities that are generated as individual migrants become entangled in the migration regime and enter Seoul under differential spatial and temporal conditions. Rather than discrepancy based on some kind of essential difference, the combination of these three shows that the observable differences in migrant lives relate to the ways their internal formations are articulated through the migration regime into everyday life in Seoul.

3.1   Migration Regime 1.0: National Development and Strategic Ambivalence

One of the most widely circulated public, official and scholarly imaginations of South Korea is of a nation that is ethnically homogenous with a shared lineage and coherent history of development. This imagination, as Shin (2006) reminds us, is one that was generated in the encroaching influence of foreign powers in the nineteenth century, in emerging western intellectual debates about racial difference, and in the claim articulated in Japanese expansionism that Japan and Korea are one and the same (see also Han, G. 2015; Lie 2014). The notion of national homogeneity and unity is one that has also been central in South Korea's narrative of social and economic development over the course of the late twentieth century (Han, G. 2015). This is apparent in early political projects that framed economic development as a shared national and ethnic objective, such as Rhee Syngman's *Ilmin chuŭi* (an ideology of one people) and Park Chung-hee's *Choguk kŭndaehwa* (modernisation of the fatherland). In these political projects and even in the more recent articulations of *segyehwa* (globalisation) a significant premium is placed on ethnic homogeneity and unity as a source of the determination and ingenuity required for successful national emergence. Set in a context of cold war geopolitics and separation from North Korea, these projects worked through a notion of South Korea as a territorially specific and contained homogenous nation state (Park 2015). The border marked a line between absolute inclusion and exclusion.

Migration has been informed in important ways by this 'modern geopolitical imagination' (Agnew 2003) of the South Korean nation. In the four decades that followed the close of the Korean War, the Korean Peninsula including South Korea was territorially constituted through a rigid border formation that controlled both inward and outward mobility. The Korean Nationality Act (1948) and the Emigration Law (1962) manifested key dimensions of this border regime in their focus on limiting and controlling the mobility of national subjects across borders. This occurred most notably through restricting access to passports to those with significant income or wealth, or through selective allocation to those studying abroad or undertaking business or government activity of national interest. In the 1960s and 1970s the Park Chung-hee administration also engaged in the strategic deployment of labour overseas, particularly to Germany and the Middle East, as part of developmental policies to generate international income from a substantial domestic labour surplus (Ling 1984). These governmental approaches to managing *emigration* reflected an underlying political rationality of building population and economy through export-oriented industrialism and for managing potential political dissent that emerged in both surplus labour and in the free circulation of citizens.

The governmental approach to the movement of non-nationals into South Korea followed a similar pattern throughout this period, although arguably there was little pressure on the state to respond to movement. The Immigration Control Law (ICL) enacted in 1963 established the basic regulatory functions of the state in relation to migration, and for almost three decades provided only very restricted access to visas or permits. Setting aside tourists, there was scope for individuals to arrive as religious workers, diplomats, business people and most notably as military in the agreements established with the UN and the USA. Setting aside these categories annual statistics suggest that there were rarely more than a few hundred other non-nationals granted permits in South Korea right up until the mid-1980s. South Korea was viewed principally as a nation of emigration, where the role of the state was to manage expectations for mobility and where necessary to capitalise on its potential as a source of income and globalising potential.

Situated in this relatively contained imagining and enacting of the nation, the presence of labour migrants in Seoul at the end of the 1980s came as some surprise. It is in 1987 that reports started to emerge of migrant workers in South Korea, initially in an article in the *Dong-a Ilbo* newspaper claiming that there were 'hundreds' of Filipina domestic helpers working undocumented in the wealthy Gangnam area (Seol 2000). At the time, there was effectively no system in place for managing or even processing the arrival of labour migrants, outside of a very small number of educators and the much larger corps of diplomats and military personnel. The South Korean government's initial attempt to regulate and normalise the mobility of migrant workers in the early 1990s came in the form of the Industrial and Technical Training Program (ITTP). Modelled after the Japanese system established in the 1980s, the ITTP effectively allowed 'trainees' to be brought into businesses for two periods of six months, during which time they could be paid 'allowances' rather than wages. As several authors have noted (Kim, W. 2004; Lim 2002; Moon 2000; Seol 2000), none of these trainees underwent training and all were effectively workers in disguise without legislated rights of minimum wage, worker protection or recourse to the law.

Irregularity was widespread during this period and was arguably an inherent part of the objectives and operation of the migration regime (Seol 2012). Throughout the 1990s as the number of labour migrants grew from a few thousand to 250,000 by the end of the decade, irregularity remained at two thirds of the total migrant population and at times reached as high as 80 percent (Lim 2002). Newly arrived migrants soon recognised or were advised through social networks that higher wages and longer tenure could be secured outside of the formalised ITTP. Government, unions and businesses did little to address this situation. In short, the migration regime represented a form of 'strategic ambivalence', 'the state knew that the number of irregular migrants was growing but did nothing to curtail it (ambivalence) simply because the country's export-oriented economy benefitted from the presence of these workers (strategic)' (Ybiernas 2013: 7).

A similarly ambivalent posture characterised the approach to the migration of English teachers and international students throughout the 1990s. In 1993 the E2 'language instructor' visa was introduced to narrow the focus of policy from the previous 'teaching and research' visa that included a wider range of occupations. The new system provided 12-month visas through employer sponsorship and eligibility was established through proof of a bachelor's degree and national citizenship. Irregularity was also common in this regime however. Many teachers would arrive on tourist visas and then search for jobs through employers or agents on the ground in Seoul and other major cities. Records are sparse but the unregulated character of the industry is revealed in guidebooks from the time that described in detail the pros and cons, ins and outs of how to be a 'cowboy teacher' (work on a tourist visa), or what was involved in making a 'visa run' to Japan or Taiwan to renew tourist status (Luxner 2004; Specht & Freeborne 1996). 'Private tuition' was also an illegal but common practice in the 1990s, providing an avenue for extra non-taxed and relatively lucrative income through networks of teachers, recruiters and parents (Lartigue 2000).

There were very few international students in South Korea prior to the twenty-first century. What student mobility did take place largely occurred under the auspices of governmental programmes for promoting South Korea, in area studies programmes established by the Japanese and American governments, or as part of developmental schemes such as the 'Look East' programme that sent Malaysian students to South Korea and Japan from the late 1980s. At the time student visas only allowed migrants to study and were limited to the duration of a course of study, there was no prospect of remaining longer or undertaking any form of employment. As much as there was little desire to actively promote educational migration there was also little effort on the part of the government, or indeed universities and other associated actors to manage the presence and practices of students or to view them as potential residents in the way that has become more common over the last ten years (Shin & Choi 2015).

Situated in relation to the narratives of ethnic homogeneity, national development and the modern geopolitical imagination, it is perhaps unsurprising that there was relative ambivalence to forms of migration when they emerged during the 1980s and 1990s. Viewed as beyond the ethnically defined nation, migrants were largely invisible to national policy makers and the public more widely. When recognised they were viewed as necessarily temporary subjects for whom little concern for rights and protections needed to be undertaken – they could not be Korean and so did not register within imaginings of 'the Korean people' (Lie 2014). The 'strategic ambivalence' (Ybiernas 2013) that characterised the early attempts to regulate mobility is indicative of this position. They represented a scenario where the border and bordering practices were undertaken only as a form of exclusion and control, an exercise in 'differential exclusion' (Castles 1995: 294) that sought to reinforce the sanctity of the nation as an ethnically demarcated socio-spatial assemblage.

3.2   Migration Regime 2.0: Managed Mobility

Over the course of the late 1990s and in particular during the first decade of the twenty-first century there has been a reconfiguration of the migration regime in South Korea and in the process to the operation of the border in relation to mobility more generally (Lim 2003; Seol 2012). Increasingly, the state has started to view the arrival and presence of all kinds of migrants as an area of economic development, has set and managed new conditions for migrants, has begun to work with non-state actors like business associations, other national governments, NGOs and universities to promote migration, and in the process has injected considerable stratification to migration processes that manifests in greater discrepancy between migrant lives. These changes have also been connected to considerable changes in the number and nationality of people migrating as workers, students and teachers (see Figure 3.1 and Table 3.1). From the 'strategic

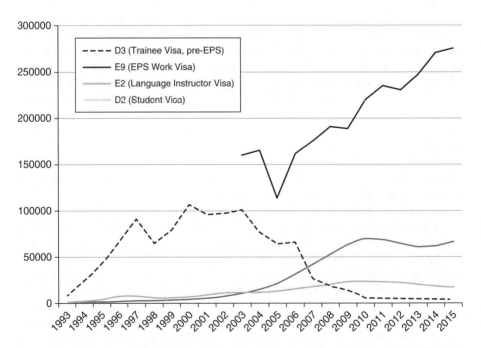

Figure 3.1   Change in number of sojourning foreigners on select visas: D3 (pre-EPS Trainee Visa); E9 (EPS Work Visa); E2 (Language Instructor Visa); D2 (Student Visa).

Table 3.1   Top nationalities of select visa categories: E9 (EPS Work Visa); E2 (Language Instructor Visa); D2 (Student Visa). Data source: Korea Immigration Service (2017).

D2 (Student Visa)		E9 (EPS Work Visa)		E2 (Language Instructor Visa)	
Nationality	Number	Nationality	Number	Nationality	Number
China	40,885	Vietnam	44,154	USA	8,183
Vietnam	4,028	Cambodia	35,409	Canada	2,519
Mongolia	2,537	Indonesia	33,793	UK	1,721
Japan	1,548	Nepal	25,761	South Africa	1,230
Pakistan	1,054	Philippines	25,503	Ireland	404
India	972	Sri Lanka	24,175	New Zealand	307
USA	949	Thailand	23,732	Australia	287
Others	14,361	Others	63,515	Others	1,493

ambivalence' that characterised early responses to migration, state and para-state practices now work through forms of *differential inclusion* (Mezzadra & Neilson 2013), a series of internal borders that articulate with wider national and continental vistas for managing mobility.

The development of new approaches to migration over the last two decades has necessarily meant that migrants are accorded much more carefully configured and differently constituted rights in their mobility to and presence within South Korea. These include the establishment of a major government-to-government circular labour migration regime, the increasing role of state agencies in the recruitment, management and employment of language instructors, and the policing of international student mobility as a tool for twenty-first-century globalisation. Each of these shifts are indicative of more general trends in international migration in recent years from a focus on exclusion and border control to an increasing emphasis on 'migration management' (Geiger & Pécoud 2013; Georgi 2010; Kalm 2010; Mezzadra & Neilson 2013). Whilst diffuse in its articulations, the discourses of migration management emphasise (Kalm 2010): the 'quality' of individual migrants; reduction in irregular migration; governance of recruitment processes; bilateral or multilateral agreements; and an emphasis on 'win-win-win' scenarios for states and migrants. Migration management is, then, a political project to shape the presence, practices, attributes and tenure of migrants. It works through the establishment of systems for 'the classification of individuals and groups according to principles of perceived threats and risks,' (Shamir 2005: 200) as well as benefits and potential and then a subsequent determination of conditions of migration.

As Seol and Skrentny (2009: 582) argue, in South Korea and East Asia more generally, migration management has largely taken the form of 'migration without settlement' where the emphasis is on temporary labour market incorporation at the expense of other possibilities for inclusion. Nonetheless, as the discussion of regulating migration in Sections 3.2.1–3.3.3 suggests, there are variations in the ways in which migrants who are workers, teachers and students are accommodated and managed during their time in South Korea. Each are subject to 'friction' (Tsing 2005; Zhang, Lu & Yeoh 2015) in their mobility and everyday lives, but the specific politics of differential inclusion and exclusion are not experienced evenly. Rather than one process of closure and containment, it would be better to think of 'several different intersecting regimes of mobility that normalise the movements of some travellers while criminalising and entrapping the ventures of others' (Glick Schiller & Salazar 2013: 189). Through border crossing migrants encounter processes of subjectification in which they are expected to participate only in certain predetermined economic sectors and where governmental technologies seek to foreclose possibilities for other kinds of social and cultural interpenetration (Yeoh 2006).

3.2.1 Labour migration

The most high-profile example of the increasingly managerial approach to migration in South Korea is the establishment of the Employment Permit System (EPS) in 2003. Replacing the ITTP the EPS established a substantial reconfiguration of governmental approaches to labour migration. Under this new system, migrants were granted work visas for up to three years (now increased to four years and ten months) and were accorded a range of basic rights: minimum wage, housing standards, access to health insurance, right to unionise and recourse to the law as workers rather than as trainees. The EPS came into force through the Employment of Foreign Workers Act (EFWA) that was passed following years of activism by migrants and NGOs (Lim 2003) that sought to reveal the substantial abuses and exploitation that had taken place under the ITTP. In enacting the EFWA former President Roh Moo-hyun (2003–2007), himself previously a human rights lawyer, described the humanitarian imperative:

> As responsibility and rights are inseparable, the nation, joining the ranks of advanced countries and the UN human rights conventions, should hold up labor policies meeting the international norms and standards not only in name, but in reality.

The political rationality that frames the establishment of the EPS was not humanitarian alone however. Rather, it is also clear that the reconfiguration of the migration regime also related to an ongoing desire to maintain and even enhance the flow of workers into key labour sectors while also protecting Korea's reputation as a global citizen (Gray 2007). Across all three of these imperatives the establishment of a highly formalised and seemingly transparent regime became a high priority, to address human rights concerns, to create more efficient flows of workers, and to establish a reputation for 'good governance' generally but also with specific regard to its Asian neighbours.

The EPS includes a number of important 'rights' for workers, around wages, housing and disputes that should not be underestimated, not least recognition of their presence as workers rather than trainees or irregular migrants. The regime remains, however, principally oriented towards improving the 'quality' of migrant workers without allowing them to gain a long-term social foothold in Korean society or perhaps more importantly not allowing unfettered mobility into, out of and through national territory. It has established a balance between having relatively skilled workers and limiting the ability for migrants to demand higher wages (Kim, A.E. 2009). Within this regime, migrants are subject to an extended or 'permanent temporariness' (Collins 2012); they may now spend nearly five years in South Korea, with all the personal and social development that occurs across that time horizon, but may not transition to other kinds of visas, start a family or remain permanently in a documented status. The insistence of the state on this temporariness is best summed up in the decision to limit the extended period of EPS workers to four years and ten months, only two months short of the period required for eligibility to apply for permanent residence.

The EPS is managed by the Ministry of Employment and Labor (MEL) and operates through agreements negotiated directly with the governments of migrant sending countries, all currently in Asia. Workers are recruited for five predefined low-skill sectors: manufacturing, agriculture, construction, fisheries and services. Employers in these sectors must be unsuccessful in recruiting Korean workers for seven days before they can apply to hire non-Korean workers. The designation of sending countries and the yearly quota of workers is determined annually by the Foreign Workforce Policy Committee (FWPC):

> … after considering which countries Korean employers favor, how transparent and efficient workforce-sending procedures are in sending countries, whether migrant workers are willing to return to their home country or not, and if there is a probability for workers to leave their workplace without permission (KoILaF, 2007: 3).

After applying through a local institution in the sending country, candidates sit the Test of Proficiency in Korean (TOPIK) and are measured on the basis of skills and experience; local institutions prepare a list of candidates for MEL, which then presents candidates to employers seeking workers. Prior to departure and upon arrival workers must undergo job training, medical check-ups and training for life in South Korea. Successful applicants are granted an E9 visa that permits them to work in South Korea for the employer who has selected them after which time they must return to their home country but may return for another stint after six months; except in special circumstances E9 visa holders cannot change employers.

The current approach to managing migration in South Korea, particularly in the EPS, seeks to regularise the movement and presence of foreign labour. Unlike the 'trainee system' that predated it the EPS includes specific standards for minimum wages, insurance, labour rights and recourse to the law that demonstrably improve the rights of migrant workers (Seol 2012). Although characterised by improvements, however, the EPS still reflects the consistent framing of migrant workers as problematic subjects. They are associated with deviant or criminal behaviour, at risk of becoming undocumented (Kim, S. 2012) or viewed as 'oppressed', 'victims' and 'to be saved' in ways that reinforce their position 'at the bottom of the migrant hierarchy' (Kim, N. 2012: 108). These discourses reflect broader representations of ethno-national difference and associations between the backgrounds of migrants and levels of development and sophistication (Kim, S. 2012). Workers are hence seen as in need of both restriction in terms of tenure and activities as well as protection from exploitation and abuse. In order to achieve this the EPS also manages migrants through stringent arrival procedures, training and orientation programmes, support services and an annual survey of worker circumstances. These technologies, while often providing useful services for some migrants, also work to maintain the temporariness of migrant presence, to allow for circumscribed work activities while foreclosing the possibility of longer-term residence.

3.2.2   English teachers

The migration of English teachers to South Korea has also undergone significant reconfigurations since the mid-1990s that have facilitated an increase in the number of foreign English teachers while also managing their presence and prospects in the nation. The first indications of a more concerted focus on recruiting foreign English teachers came in the introduction of the English Program in Korea (EPIK) in 1995. EPIK mirrored the equivalent Japanese English Teaching (JET) programme by inserting the government as a key actor in teacher mobility that has been dominated by private industry. It was situated in relation to the growing focus on *segyehwa* under the Kim Young-sam administration and was explicitly framed as a mechanism for 'reinforcing foreign language education' and 'reinforcing globalization education' (Jeon 2009), although it has more recently also been articulated as a mechanism for 'encouraging cultural awareness' and to 'enhance Korea's image abroad' (NIIED 2016). Initially a relatively small scheme with 54 teachers in 1995, EPIK recruited 3,477 teachers in 2012 (Yoon, S. 2014). Similar initiatives have been developed locally in the Seoul metropolitan region like the English Teachers in Seoul (ETIS) programme, established in 2003, and the Gyeonggi English Program in Korea (GEPIK) that began in 2004. Like EPIK these programmes place teachers in public schools as 'assistant teachers' alongside Korean English teachers, their role is to increase the use of English and the immersion of students in English language norms.

These programmes are situated within a wider communicative turn in the teaching of English in South Korea. In 1994 the Ministry of Education and Human Resource Development (MoE) required all middle schools to teach English and from 1997 this was extended to Grade 3 students in elementary schools. More significantly, the 7th National Curriculum of Korea (2000) shifted the focus of English teaching from grammatical and technical language knowledge to communicative competence, which was later extended through the requirement in 2001 that English be taught through English medium only. This communicative shift altered the position of Korean teachers of English and increased the pressure on recruiting native speakers who were presumed to be more competent in communication, even if not in the technical features of language. These changes manifested most obviously in an increasing number of native English speakers entering South Korea to teach in public schools but also in a now booming private education market. In 2000 there were 6,414 migrants holding the E2 visa, increasing to 8,388 in 2001 and 10,864 in 2002, and by 2010 the number of 'language instructor' visa holders entering South Korea numbered 23,317.[1] While the number of migrants working in public system programmes like EPIK, ETIS, GEPIK or Teach and Learn in Korea (TaLK) grew during this period, the vast majority of teachers are recruited for private sector academies either directly by employers or through recruiting agencies.

Unlike workers arriving through the EPS, the migration of language instructors in South Korea is managed by the Korea Immigration Service (KIS), which forms part of the Ministry of Justice (MoJ). This notable division of governmental labour reflects the ideological framing of English teachers as travel and work sojourners rather than simply as labour power. To work in either private or public schools individuals must hold an E2 'language instructor' visa. To qualify for this visa applicants must be 'natives of the country whose mother language is the one they are teaching', which for English is defined by KIS as Australia, Canada, Ireland, New Zealand, South Africa, the UK and the USA. Applicants must possess a bachelor's degree, but need no formal training in language instruction. In addition to criminal record checks, migrants have between 2007 and 2017 been required to undergo drug and HIV tests prior to being granted an E2 visa – standards that are not required for any other migrant group entering South Korea.[2] In July 2017 the government announced that this requirement would be scrapped after it was ruled a form of racial discrimination, although prospective teachers still have to be tested for drugs and syphilis. In the public sector teachers often undergo an orientation or training programme prior to starting teaching but in private sector jobs this is much more uncommon and migrants often begin teaching shortly after arrival in South Korea. The E2 visa is granted for 12 months and may be renewed indefinitely with employer sponsorship. E2 visa holders may change workplaces once their employment contract has been terminated (end of contract, resignation or dismissal). E2 visa holders are also permitted to reapply for visas during short visits to neighbouring countries and if they become eligible can transition onto professional work visas and even permanent residency in some instances.

This approach to managing the migration of English teachers reflects conceptions of English education in South Korea and in particular the idealisation of the native speaker as both the representation of perfect English and the most credible teacher of English (Shin, H. 2007). Race forms a critical component of these idealisations with English often framed as 'the property of white American English speakers' (Ahn 2014: 212). Accordingly, there has been a gradual shift away from English taught by Korean English teachers with knowledge in linguistics towards an increasing presence of native speakers as teachers. This has manifest first in the massive market for private English academies but more recently in public schools where policies aim to secure at least one native English teacher per school (Jeon 2009). Implicit also is a framing of English teachers as fitting within particular stages in their life course, young recent graduates without families for whom time abroad will not be considered an impediment. These differences then manifest in hierarchies of hiring that vary by employer but generally favour younger white North American teachers, followed subsequently by other nationalities, ethnic minorities and older teachers. At the same time, this alignment of youth, independence and western culture has also however fed into moral panics about teachers and their influence on young Koreans that the government has sought to address publicly through forms of drug and health testing. While desired, then, English teachers have also been seen as in need of management.

3.2.3   International students

Unlike both EPS workers and language instructors, international students do not form part of orthodox conceptions of labour migration (Brooks & Waters 2011). Nonetheless, the increasing number of students in South Korea is situated within wider discourses about demographic renewal and about the human capital circulations associated with migration (Shin & Choi 2015). As has already been noted, there were only a few thousand international students in South Korea at the end of the 1990s, most of whom were concentrated in the discipline of Korean studies or government sponsored programmes. By contrast, over the first decade of the twenty-first century South Korea became the fastest growing destination for international students globally (OECD 2012), and in turn international students became the fastest growing migrant group in South Korea over the last 15 years, increasing in number to 20,347 in 2005 and 69,600 in 2010. Although students came from 168 different nations, students from Asia represent over 90% of all students with Chinese students (including Korean-Chinese) representing 55,008 or 64% of all students.

The rapid growth of international students articulates with important shifts in the higher education sector in South Korea over the last two decades. First, building on the *segyehwa* movement there has been a much greater focus on developing 'knowledge economy' initiatives that might help transition the South Korean economy from its export industrial developmental model to a focus on post-industrial services and technologies. Initially this took the form of increased research funding for major institutions, like the Brain Korea 21 and World Class Universities initiatives, which Seoul National University and large Seoul-based private universities like Korea University and Yonsei University received the lion's share. More widely, there was an increased emphasis on international reputation both measured in terms of standardised ranking mechanisms but also around the desirability of leading institutions in global flows of students and faculty (Collins & Park 2016). With its eye on diversifying student populations and capitalising on the growing profile of South Korea in Asia, the MoE established the 'Study Korea Project' in 2004. The project identifies three priorities: (1) to link advanced and developing countries to support human resource development of developing countries; (2) to globalise and strengthen the competitiveness of Korean education; and (3) to promote Korea's image as a destination in ways that capitalise on the success of the 'Korean wave' of TV dramas, popular music and film. These priorities are promoted through the provision of government scholarships, recruitment strategies, and incorporating study into the broader publicity and marketing of Korea internationally.

The recruitment of international students is overseen by KIS in relation to the Study Korea Project and the institutions that students enrol in. To be eligible applicants must be enrolling in a higher education institution. Applicants must establish that they have appropriate entry qualifications and that they have the financial resources to support themselves. The management of student migration, in this respect, is also tied up with the activities of universities in terms of advertising programmes, recruitment, enrolment and administration and latterly in terms of the incorporation of students into life in South Korea. While students are migrants and experience certain kinds of restrictions on their mobility, these are not like those that shape the migration of EPS workers or language instructors. Unlike migrant workers and English teachers, students do not provide a criminal record or health check. Student visas can be issued for up to five years, with allowance for multiple entries and no restriction on changing programmes of study once a visa is issued. In addition to study, international students can work 20 hours per week for undergraduates or 30 hours for postgraduates. After graduation students may apply for a 'Job Seeking' visa to look for employment or 'Professional Employment' visa if they have a job offer, which will provide the scope to work and live in South Korea indefinitely. More than just workers filling particular occupational niches then, students are understood for their future potential for the Korean economy and society (Shin & Choi 2015).

International students occupy a much more contested position than the rather more clear-cut demarcation of 'migrant workers' and 'English teachers'. On the one hand, the recruitment of international students is framed within state discourses of needing to secure 'global talent' (Shin & Choi 2015) in a context of demographic decline and South Korea's increasingly transnational economic activities. Within such discourses international students serve as potential bridge-heads into new markets and provide the diversity of skilled labour that Korean companies perceive as necessary in the twenty-first century (Collins 2014a). Universities and the higher education sector more generally, have also viewed international students as an avenue for increasing global reputation, for diversi-fying campus life and learning environments and advancing Korean higher education (Collins & Park 2016). In both cases, international students are framed as futural subjects, holding potential to alter the trajectory of institutions, business and national society and economy in an era when diversity is at a premium. At the same time, however, international students are also viewed as problematic sub-jects. Their presence is bemoaned as a cause of dropping standards at universities, and their demand for English Medium Instruction (EMI) has led to claims that 'Korean universities have lost their cultural uniqueness and thus have become indistinguishable from universities in other countries' (Cho & Palmer 2013: 302). Moreover, claims also circulate about the legitimacy of international students, raising questions about whether student mobility is a backdoor to labour migration that should be more appropriately governed through the EPS or other migration regimes. The notion of students as workers operates through reference to national origins and the predominance of Asian students in Korea, connecting their trajectories with those of migrant workers from the same backgrounds (Kim, S. 2012). International students, then, inhabit a rather contested socio-political position that reflects the intersection of perceptions about national origins with state generated rhetoric about students as bearers of global talent.

3.3   Biographies of Desiring-Migration

As regulatory and bureaucratic functions, these regimes and discourses operate through abstract spatialities where migrant bodies are constituted as legible in terms of a range of demographic and human capital dimensions. Beyond this reified world, migration regimes also intersect with and are constituted through the material lives and mobilities of migrants themselves. They become interlinked with other dimensions of migration that start not in the desire of states to exclude, control or manage movement (Doty 2003) but rather in the desire for mobility of individual subjects and the manner in which this becomes articulated with certain opportunities while also blocked or closed off from other opportunities (Collins et al. 2014). In order to unpack the articulation of migration regimes with the mobilities of migrants themselves I move now to discuss desiring-migration through the lives and narra-tives of three individuals who were interviewed as part of this project. The narratives of Nadia, Nonoy and Jiaying discussed here hint at the emergence of these discrepancies in terms of the timing and prospects of migration.

A biographical approach is particularly valuable for exploring the unfolding of variant forms of desiring-migration and its articulation with extant migration regimes. Rather than viewing migration as an irreversible decision undertaken by an atomised individual at a singular point in time, a focus on the biographies of migration reveals the important spatial and temporal contexts of generating migration as a desirable undertaking (Christou 2009; Lawson 2000). Individual encounters with mobility, and its mediation through migration regimes, need to be contextualised within an understanding of what individuals bring from their pasts into enactments and experiences of movement (Rogaly 2015). Focusing on biographical approaches also provides opportunities to explore both unexpected similarities in migrant biographies and discrepant outcomes as individuals interact with the uneven terrain of migration.

The discussion of Nadia's, Nonoy's and Jiaying's narratives below focuses on drawing out some of the very particular ways in which desire emerges as a force in migratory processes, how it is enabled through connections, impeded through blockages, and reworked through migration regimes that entangle migrants into wider patterns of mobility and migration management. The narratives presented here, hence, are not intended to be indicative examples of the 'reasons' or 'patterns' of migrant mobility to South Korea. Rather, these narratives offer us a lens through which we might conceptualise the wide range of forces at work in making migration possible, channelling migrants in particular directions, and generating always unexpected transformations in individual and collective lives.

3.3.1   Nadia

Nadia grew up in a small town in Cornwall, a relatively sparsely populated part of Southwest England. She described the places she grew up as 'kind of boring', characterised by beaches and the lack of people except in the summer season when there is an influx of tourists. After finishing secondary school Nadia went on to study technical theatre at a university in London, which 'was really cool, I really enjoyed being in the city and meeting so many people'. Graduation from university posed more challenges, however, and like many migrants in this research Nadia was only able to find temporary work that did not align with her educational achievements:

> A lot of my friends have left [Cornwall]. Yeah, they've moved to different parts. I mean, if you want a really good career, then you know, it's quite hard to get it there cos' it's such a tourist destination, you know? In the winter there's not a lot of jobs you can do and actually work in an office or that kind of thing. … I did some travelling for a while and I was only working in bars and things – it's so hard to get a job in theatre in London. And you know, I just was getting stressed, I was like 'What can I do? What can I do?' Like I wasn't earning much money, so then I thought alright, I'm gonna go somewhere.

Stories of this kind are common amongst foreign English teachers in South Korea, although they also resonate in interesting ways with the accounts of migrants arriving through the EPS. Interviewees described an interrupted transition from education, whether secondary schooling or higher education, to work and in particular into fulfilling and financially stable careers. These narratives are indicative of the way that neo-liberalisation of social and economic life has generated heightened expectations around education and individual achievements for young people in particular while also narrowing the scope for achieving these aspirations through stable employment (Furlong 2014). In Nadia's case, this situation articulates with the rapid massification of higher education that has taken place in recent decades in the UK and other western Anglophone countries and the flow-on effects for 'qualification inflation' and graduate underemployment. For many, this predicament was exacerbated by student loans or other kinds of debt that compounded anxiety about future prospects with the 'stress' of day-to-day living costs and income in the expensive metropolises of the UK, North America and Australasia.

Nadia's response to this crisis was to abandon her plans to remain in London and the UK and to explore other options. She had at this stage not considered teaching English or going to South Korea at all. Indeed, she recalled that she 'didn't know anything about South Korea'. It was instead through searching on job websites and generally on the internet that the prospect of going to teach English and to South Korea first emerged:

> I was just browsing the Internet and looking at jobs for overseas, I wasn't looking for Korea in particular – just anywhere. Just to see what's available and what kind of thing I could do with my qualification, and I was particularly interested in Asia because when I travelled, I spent about six weeks in Thailand and I really loved Thailand and you know, kind of got me interested in Asian culture. So I decided to maybe look into China – it was my first option, and I got in touch with a recruiter, and he kind of persuaded me to consider South Korea because he said many people don't really know about South Korea. ... So I did some research and I thought 'Oh it looks really pretty' [laughs] Uh, yeah. I decided South Korea.

South Korea emerges at a particular juncture for Nadia, then, where she is addressing present concerns and challenges, but the idea of mobility also demonstrates the importance of being linked into the circuits of information about migration and its possibilities. The imaginative dimension of migration as desirable is prominent here, although for Nadia and many other English teachers South Korea is subsumed within a wider geo-cultural framing of 'Asian culture' that links tourist encounters in Thailand to China's rising presence on the global stage and South Korea as an alternative destination. Desire is important here, not as a fully formed force, but one that works through Nadia's extant interest in 'Asian culture', the recruiters redirection of her focus away from China and in her virtual exploration of South Korea through the internet that generates an imaginative geography of beauty and intrigue.

This account of finding and being attracted to South Korea also speaks to the key role of other infrastructures and actors in these migratory processes (Xiang & Lindquist 2014) – here the internet and various job search engines are key as well as the 'recruiter' (a key figure in contemporary migration studies – Lindquist 2015). Indeed, the latter not only provides information about teaching opportunities but also creates a blockage in relation to China and draws attention to South Korea by emphasising the desirability of the latter as an alternative destination. In the narratives of other migrants, recruiters were also prominent but many spoke about the importance of friends as conduits of knowledge as well as intermediaries who sometimes made arrangements for jobs, accommodations and provided visa information.

In the case of English teachers like Nadia these arrangements can emerge very quickly and move migrants along a migratory path at a rapid pace. She described how the recruiter became very insistent about the timing of her arrival, 'they kept giving me offers like "Oh, you can start now, start now"'. Nadia held her ground about her preferred arrival and eventually came to South Korea about three months after first discovering this opportunity. This pressure was common for English teachers, some of whom were literally offered jobs within hours of sending emails to recruiters and found themselves in a new country with a visa and a new job within a matter of weeks. This was not an experience shared by other migrants in this research, particularly those arriving through the EPS whose experience of migration was characterised much more by suspension, pauses and waiting.

3.3.2 Nonoy

Like Nadia, Nonoy's narrative of migration to South Korea is situated within particular geographical contexts that shaped his outlook on mobility and its prospects. Nonoy came from a family and community where migration overseas was the norm. He grew up as the eldest of three children, born to a Spanish mother from Barcelona and a Filipino father, who lived across four different provincial areas of Luzon, the largest and most populated island in the Philippines. His aunties raised him as his mother worked in different regions as a clerk in a Mayor's office, a teacher, a farm labourer and 'managing things'. Nonoy's father spent most of his childhood working in Saudi Arabia as a contract construction worker. All of Nonoy's cousins have lived and worked abroad and his younger sister is currently working abroad as a first-aid assistant in shipping companies.

After completing high school, Nonoy went on to graduate with a bachelor's degree in computer information systems from a university in the northern part of Luzon. Initially following graduation Nonoy started working as an IT teacher but he 'had itchy feet' and started thinking about undertaking a nursing degree so that he could learn more and find other opportunities. Although he grew up in a mobile family, migration had not been something that Nonoy had thought that much about until a cousin of his dragged him along to a recruiter.

> I was going to study nursing. But then, my cousin, he said, 'I want to study again, while working', since, I was working as a teacher right. But I got my forms and my cousin from Hong Kong, who's also a teacher also a graduate said, 'let's go to the recruiter'. See, his siblings had all been to Korea before. So, it was just us among the cousins who were left here, so I said 'Sure! I'll come along'.

Flexibility and mobility are key components of Nonoy's narrative and more generally of the wider articulation of young people's lives into spaces of migration. As Lukacs (2015) argues, young people are a pivotal component of the neo-liberalisation of Asian societies, both a demographic excess that are excluded from historical renderings of the developmental state and yet also valorised for their vitality and energy and their capacity to change directions and remain flexible. Nonoy's shift from information systems, to teaching IT to potentially nursing and as we shall see to migration to South Korea epitomises this flexibility, a desire for something different as he suggests but also something that can draw young people into precarious situations.

Nonoy made it clear that he 'didn't really have plans to push through with' applying for migration but was just humouring his cousin who had these plans to go abroad. The encounter with the recruiter reshaped Nonoy's desires however, and within a short period he had shifted from planning to pay his tuition fees for nursing to undertaking tests and exams to go to South Korea.

> At the recruiter's, she said, 'Would you like to go to Taiwan?... Because you're in already'. All they look, all they look for in Taiwan is just the appearance, anyway. They want people who look tall and then decent-looking, good-looking. That's what they... so I said, 'Well!' But my cousin said uh, the salaries were quite low. Then [the recruiter] said 'what's booming now is Korea'.

Timing is critical in this narrative, in the way in which Nonoy has been presented with this opportunity but also in the chronology of South Korea's evolving migration regime. The year is 2004 just as the major reforms in the labour migration system in South Korea were getting underway and the EPS was being established as a new scheme for migrants – there was a minimum wage, backing from the government and a flood of South Korean popular culture and tourists in the Philippines. South Korea, as the recruiter put it, is 'booming now' and Nonoy was presented with an opportunity to travel overseas, work in a new environment and try something very different from what he had done to date. Desire lurks in such encounters, not because individuals suddenly change their minds, but because desire is a force that draws subjects into wider social assemblages that generate possibilities that were either unknown or shrouded previously – individuals themselves become part of a migration assemblage (Rubinov 2014).

After a week Nonoy received a call back from the recruiter telling him, 'you're in for the interview'. He had to make a tough and quick decision – he was due to pay his tuition fees for nursing school that day and also was required to attend a medical examination, without which he would not be able to go forward for migrant selection. He did the medical but still felt that he was always going to go ahead with his studies, that was 'the only thing on my mind'. It was his parents, who themselves carry their own histories of migration and its prospects, who added the extra motivation:

> I talked to my mom and dad. I said, 'well. I might be going abroad ... I would like to study again though'. I really wanted to study, and they knew it, that I wanted to study again, until now. And then [they] said, 'just try it. Just give it a try', they said. '*You don't know your luck* if you'd like to give it a try, since your papa went abroad in the past too'. 'As for me ... I worked in another country too', he said. 'Try it out'. Then he said 'just try it' ... and then luck comes ... I passed the medical exam. So that's when I started training.

While there are important similarities between Nadia and Nonoy's encounter with South Korea as a place of potential migration, their articulation into the migration regime differed considerably. Unlike Nadia, Nonoy was not pressured to depart immediately nor offered easy access to a visa and it was eventually one year before he would depart for Seoul. During this time Nonoy had to complete medical exams, language training and tests, and prepare wide ranging documentation to be submitted and processed through Filipino and South Korean government authorities before being sent to potential employers who would select Nonoy rather than vice versa (as was the case for Nadia). To pass the time before departure Nonoy kept working as a teacher but he also sought out a Korean volunteer at the university he worked at and through her learnt some Korean phrases, tried some Korean food – including a month of eating Kimchi every day – and watched lots of dramas, films and music to learn more about South Korea. This process of preparation engendered both excitement and anxiety about what might lie ahead, feelings or affects that we can understand as common components of desire, especially when it is held in temporal suspension in this way.

3.3.3 *Jiaying*

Jiaying grew up and completed all her schooling in a smaller city in Henan Province in the central area of China. She described her family life as 'nothing very extraordinary', her mother had an undergraduate degree and worked in a bank and her father, who had completed high school, worked in a tobacco factory. Jiaying described an upbringing where expectations were not particularly high but where her parents allowed her and her brother to do 'whatever we want to do, they'd let us go ahead and do it'. As a family she travelled around China when she was young but had never travelled overseas before coming to South Korea, mainly because of the complexity of getting visas on a Chinese passport.

When Jiaying completed high school she was considering various options for university study but had been looking at overseas options because this was the normal thing to do and because her aunt was based in South Korea this emerged as a prominent option.

> When I graduated from high school, my family, these days there are many people who go overseas to study, so my family said that since my auntie is here, and as a girl, it'd be better not to go to a country which is too far away, better to go to a country which is nearer. So I chose to come to Korea because my auntie is here studying. This is one of the more major reasons because at that time, when I first came over, I had just graduated from high school and was still not very 'sensible', she would be able to take care of me. In addition, for Japan, the reason why I did not go to Japan is because, you would know that the Japanese do not really like the Chinese to go to Japan.

Unlike both Nadia and Nonoy then, Jiaying's migration was relatively prefigured and not characterised by the sudden emergence of a migratory opportunity or the prospect of South Korea. Rather, like other international students Jiaying had been thinking about overseas study during her schooling, her parents, friends, teachers and others had been suggesting various places as possibilities and she had developed a detailed imaginative geography of what was possible through migration (Salazar 2011). Within this imaginative geography South Korea offers certain opportunities while other places like Japan are blocked because of perceptions about the place of Chinese in Japan. When it came to South Korea it was the presence of her aunt that was pivotal but Jiaying had also been exposed to the television dramas and music of the Korean wave. Like other migrants who came as international students and through the EPS, exposure to the Korean wave provided a web of intelligibility for Jiaying, a set of images that triggered desire for migration, but that were often found lacking in relation to lived experience in Seoul: 'it looked quite good but when I came here ... it's not as good as China'.

Jiaying's migration is also situated in a context where expectations around education as an imperative for youth have become increasingly pronounced. Over the last two decades, the proportion of young people undertaking international education in China has grown from some 120,000 in 1998 to over 800,000 in 2016 as the massification of higher education has created more pressure on skilled positions and higher incomes have allowed for some flexibility in study options. As Xiang and Shen (2009: 516) note much of this mobility is generated in 'anxiety about the future, and the rapid social stratification in China is a fundamental driving force for of this anxiety'. Of course, these patterns are more widespread and certainly within Asia there has been a particular growth in student migration as a response to changing conditions for young people at home (Collins & Ho 2014). The emergence of seemingly lower cost and proximate destinations like South Korea has meant that students who might not otherwise be mobile are now presented opportunities to be 'one step ahead' and to access the valorised cultural capital of an overseas degree.

In order to arrange her migration to South Korea, Jiaying utilised the services of an education agent who could apply for visas, make arrangements for enrolment and connect her with other Chinese who were migrating as international students. As she described it, 'the screening process is quite strict' for Chinese nationals because study visas are sometimes used to seek work opportunities. There is in this context a need to carefully negotiate the migration regime, which for Jiaying meant she needed to work with an intermediary:

> [To arrange migration yourself] you'd have to prepare especially more information and will have to prepare a lot of deposit money, security deposit money. Its more troublesome. [Agents] have links with Korea. In the beginning, we wanted to do it ourselves when it came to learning Korean, but on your own, it's difficult. You won't be able to go to Seoul National University and Korea University. To go to these schools would be more difficult if you apply on your own, but through these agents, as they have connections with the school, it'd be easier to come over. But you would have to pay a certain amount of administrative fees.

Jiaying's encounter with the migration regime in South Korea reveals the importance of migrants establishing appropriate credentials as a particular kind of migrant, in this case as an international student rather than as a worker who would need to migrate through the EPS. Intermediaries are critical to this negotiation, providing information and guidance and then articulating information provided by migrants through their 'links' and 'connections' not only in the state agencies involved in migration but also with universities as institutions who expect applicants to present themselves in particular kinds of ways. As Jiaying indicates, not utilising an agent would make enrolling in the best universities like SNU and KU much more difficult. Her agent arranged for her to study Korean language for twelve months at KU before supporting her enrolment in SNU in a business degree.

The temporalities of international student migration to South Korea often follow calendrical rhythms around university application deadlines and semesterisation. For some students like Jiaying their study plans require preparation in Korean language in advance, often in South Korea itself, that occurs before they formerly start their university education. Unlike migrants who came as English teachers or through the EPS then, these temporalities are largely predictable in advance, they are set out on webpages, informed by agents and match expectations of individuals already predisposed to higher education systems. There is neither an intense pressure to apply for visas and depart rapidly that was narrated by English teachers nor the kind of suspension and waiting that characterised migration through the EPS.

3.4   From Desiring-Migration to Discrepant Lives

The narratives of Nadia, Nonoy and Jiaying speak to the ways in which migration trajectories to South Korea both resonate amongst and vary across the lives of English teachers, international students and migrant workers. Like other participants in this study, their narratives speak to the ways in which migration cannot be understood through the logic of a decision that is undertaken through rational calculation at a singular point in time. Rather, their movement is articulated through a process of encountering the prospect of South Korea as a place to go, characterised by certain opportunities and constraints. Their mobilities are also clearly socially distributed. Rather than only revolving around atomised individual decisions, migration is revealed here as a process that hinges off actors like agents, is situated within social norms around mobility, as well as expectations and desires amongst young people. Through these resonances however it is also possible to identify variations or discrepancies that are shaped by their articulation into the assemblage of migration. Most notably through their accounts we see discrepant temporalities, from the accelerated pressure Nadia experienced to Nonoy's year long wait for selection and Jiaying's movement through the standardised cycles of matriculation, language study and entrance. In public discourse and government policy the migration of workers, English teachers and international students have regularly been treated as separate and distinct. Labour migration through the ITTP and the reformed EPS has been articulated as a response to labour shortages, English teachers as a means to improve language acquisition and communicative competence in a global context, and international students as potential human capital for globalising universities but also as a demographic injection into skilled labour markets. The drawing apart of these narratives of generating and managing migration reveals the different status and rights given to migrants and their position within society.

The configuration of mobility through these migration regimes clearly directs migrants into particular positions within divisions of labour and life in South Korea. In the case of migrant workers and English teachers these distinctions are based upon ideological assumptions about nationality and the ability to fill labour niches based on knowledge about income levels, economic development and the cultural capital of origin countries. Although both groups are subject to pre-screening, they are accorded quite different capacities in their daily lives. While English teachers can move freely between employers, access multiple-entry visas and renew their visas indefinitely, migrant workers' sojourns are tightly regulated in ways that limit their ability to do anything except work for predetermined sojourns. International students, by contrast, are relatively lightly regulated in their migration. They are not subject to any pre-screening and are permitted to engage in a range of activities in South Korea. Family can accompany both English teachers and international students and their spouses can access employment visas; family accompaniment is explicitly precluded for migrant workers. Likewise, English teachers and international students are permitted to transition to other visa types, including Job Seeking, Professional and Residence, while migrant workers are explicitly excluded from these categories.

Placing these migrations and the variant policies deployed to manage them alongside each other also demonstrates their co-constitution as responses to the perceived challenge of modernisation in a globalising world. As the above account has shown, each area of migration has attracted increased attention from the South Korean state over recent decades that reveal the importance given to managing flows of people across borders. Together, alongside other migratory schemes like those associated with international marriage migrants (Lee, H.K. 2008), these schemes represent part of a wider migration regime that is focused on selecting, processing and placing migrants as they cross borders. There is a marked difference between this emergent regime for regulating migration and the kinds of 'strategic ambivalence' that was expressed and enacted in earlier years. These new regimes do more than simply generate absolute closure and either exclusion or inclusion. Instead, they work through different individual and group classifications, measuring potential value, skill, threat and risk. Such classifications then clearly serve as a basis for the differential encouragement of migration and the social distinctions that are made in terms of migrant rights and opportunities. This is stark in the discrepant regulation of labour mobilities of EPS workers and English teachers, but is also revealed in the relatively lightly regulated movement of international students vis-à-vis the other two groups and in particular EPS workers, many of whom come from the same nationalities.

In the chapters that follow I present more detailed accounts of the discrepancies that emerge as these and other individuals arrive in Seoul and become part of assemblages of labour and life in the global city. The narratives of Nadia, Nonoy and Jiaying hint at the emergence of these discrepancies in terms of timing and the prospects of migration but it is to the everyday lives and urban circumstances of migrants that we must turn to for further depth. As we move outwards from these lines of desire, migration and arrival into Seoul the discrepancy between lives becomes more explicit as different workers, teachers and students very literally and figuratively inhabit different spaces in the city. Accordingly, the chapters that follow prioritise the particular configurations of each group as means to reveal the different dimensions of migration's articulation with urban life, draw attention to intersections and disconnections between workers, students and teachers and to highlight the multiple subjective positions that emerge within particular migrant categories. Following on from Chapters 4, 5 and 6, then, Chapter 7 draws these lives analytically together again to explore encounters, the social assembly of urban life and the coupling and decoupling of migrant and urban futures.

Acknowledgements

Portions of this chapter have been drawn from Collins, F.L. (2016). Migration, the urban periphery, and the politics of migrant lives. *Antipode*, 48(5), 1167–1186. doi: 10.1111/anti.12255 and Collins, F.L. (2016). Labour and life in the global Asian city: the discrepant mobilities of migrant workers and English teachers in Seoul. *Journal of Ethnic & Migration Studies*, 42(14), 2309–2327.

Endnotes

1   These figures, sourced from the Korea Immigration Service, cover all 'language instructor' visas including those issued to individuals teaching languages other than English. The vast majority are however English teachers, with substantially smaller numbers even for other commonly taught languages like Chinese (1,117) and Japanese (1,079).

2   The introduction of drug and HIV tests for E2 visa holders has been controversial. The requirement was initially introduced as emergency measures to 'ease the anxiety of the citizens' following the discovery that a known paedophile, Christopher Paul Neil, had been working in an English academy in Gwangju. The measures were initially introduced for E2 visa holders, as well as E9 labour and E6 entertainment visas but the requirement was removed from the latter two following human rights complaints in 2010. The response constituted part of a wider moral panic about the growing presence of young westerners as English teachers, their behaviours and their potential impact on the attitudes of the young Koreans they most often taught (Wagner & Volkenburg 2012).

Chapter Four
Migration, the Urban Periphery and the Politics of Migrant Lives

Phuoc (Vietnam, Male, EPS worker) arrived at Incheon International Airport in the winter of 2006. Like most of the individuals in this research who come to South Korea through the EPS regime, Phuoc's experience of migration, arrival and initial settlement was very structured and something that he knew little about in advance. He had been selected after applying for the EPS while studying at a technical college near Hanoi. Prior to departure he completed necessary medical and character checks, participated in compulsory language and cultural training and waited about a year for deployment. Although he felt 'lucky' to be one of the twenty successful applicants from his college he recalled knowing very little about Korea beyond what he had seen on television, and not really knowing where he would be working or what he would be doing; that 'was chosen by the Vietnamese and Korean Ministries of Labour. I didn't know it at all'. Following arrival at Incheon, Phuoc was taken to complete a four-day compulsory training and orientation programme run by Human Resources Development Korea. He was then collected by his employer and taken to his workplace and accommodation, which were co-located in an industrial area in Ansan, Gyeonggi Province. He describes his initial reaction:

> When I first came to Korea it was not what I expected because I thought Korea was a developed country and the houses or daily activities would be modern. But unluckily, we worked in a company where the accommodation is not convenient. At the time when I first came here, the company arranged my accommodation. We lived in a container; it was just $20 \, m^2$ but six people lived in it.

Global Asian City: Migration, Desire and the Politics of Encounter in 21st Century Seoul, First Edition. Francis L. Collins.
© 2018 John Wiley & Sons Ltd. Published 2018 by John Wiley & Sons Ltd.

Having imagined South Korea in advance as a 'modern' 'developed country' and configured his anticipation of migration around this imaginative configuration, Phuoc like many other EPS workers was taken aback by the character of the place he was expected to work and live. On the one hand, this disruption of the imaginative geographies that shaped Phuoc's migration is not uncommon and reflects the processes of uprooting that are well known in the study of migrant lives (Ahmed et al. 2003). Teachers and students also regularly expressed a need to reconfigure their understanding of place as they arrived in Seoul and commented how the imaginations that formed part of desiring-migration jarred with the day-to-day reality of the metropolis. On the other hand, however, it is critical that we recognise the particularity of the disjuncture experienced by Phuoc and many other workers arriving through the EPS. Where their narratives before coming to Seoul emphasised the developmental status of South Korea and the desire to be part of a dynamic metropolis, the process of migration led them to the urban periphery, a seemingly un- or underdeveloped landscape that often fared unfavourably in comparison to their pre-migration experiences. Far from the bright lights and dynamic urban core of Seoul, Phuoc found himself housed in an overcrowded converted shipping container located in an industrial zone where his only daily contact is with other workers in and around the factory.

> I didn't feel comfortable when I came here. As I have said, those who helped me were mostly Vietnamese friends around me. There were not many neighbours, not many Koreans because they don't live in that area. [Even so] the most difficult thing I faced at that time was the language barrier. I learned Korean quite well before. But when I came here, my Korean language is nothing compared to Korean's.

Phuoc's account is indicative of the specific patterns of migration and emplacement that are experienced by individuals moving through the EPS migration regime. While the actual conditions of working and living spaces varied considerably amongst participants, the highly structured experience of arrival and the location of worker lives in the periphery of Seoul's metropolitan region was a universal feature of their narratives. Participants spoke of being organised to travel in groups from their home countries, in some cases being asked to wear uniforms and name tags, and being met upon arrival at Incheon International Airport and taken as a group through immigration formalities and then by bus to training centres for 3–4 day introductions to language, cultural norms, labour laws and workplace skills.

There is a stark contrast here with the equivalent narratives of English teachers and international students, who while also having varying circumstances were all accorded a level of freedom on arrival that EPS workers were not. Teachers, while often working in the days following arrival, invariably spoke about arriving in an apartment and exploring the local neighbourhood and students spoke of orientation programmes and social networking opportunities or campus tours. Francesca (Australia, Female, English teacher) described her feelings in the days following arrival:

> I think I was very excited to be there, it was very exciting and I was just spending my daytime, eight hours to fill in, walking around the streets, looking in the shops, very thrilled and excited, just having a lot of fun, weird experiences, I was kind of blown away and I had a lot of time to myself, so I was just exploring all by myself.

While excitement or interest in migration was common amongst participants before coming to South Korea, as the narratives of Nadia, Nonoy and Jiaying have revealed in Chapter 3, discrepancies emerge quite clearly through arrangements in the management of migrant lives in Seoul. These arrangements shape experiences of arrival, migrants do not arrive only on their own terms but rather become part of the urban fabric through the varying rules and pathways established in governmental systems. While all migrants have expectations applied to them, these conditions give rise to different capacities to direct one's own mobility in the city, from highly channelled pathways enabled by migration governance and employer practices to much greater levels of freedom to negotiate relationships with the city.

This chapter takes the urban geographies of migrant lives as its starting point to explore the politics of migration through a focus on the intersections between migration regimes and the everyday lives of migrants in cities. As Chapter 3 has demonstrated, the typical goal of such regimes is to generate, codify and regulate migration of productive subjects that have a particular position in urban life, both in terms of their social rights and capacities as well as in their geographical emplacement in the city. The contention of this chapter is that while migration regimes are largely successful in configuring the socio-legal status of migrants, the desire for control is often displaced at the level of everyday life, the ordinary and extraordinary presence and practices of migrants as urban residents. To explore this disjuncture the chapter highlights the narratives and everyday lives of individuals migrating through the EPS while also at times contrasting with the urban geographies described by teachers and students. The discussion centres in particular on the ways in which migrant lives are shaped by their place in the city focusing most clearly on positioning of workers in the urban periphery, both literally in the outskirts of the Seoul Metropolitan Region but also figuratively in their marginalised position within society and economy. Despite manifesting marginalisation the narratives of migrant workers in this chapter also point to the significance of the periphery as a space for generating different opportunities: a 'mobile commons' (Papadopoulos & Tsianos 2013) that supports migrant lives; providing invisibility as migrants seek to remain undocumented and subvert the control of migration regimes; and the ways in which migrants can actively reconfigure the periphery through tactics of recognition that materialise in the appropriation of public spaces and collective action. These practices of subversion, solidarity and recognition illustrate the political content of migration and migrant lives and point to its potential for challenging and reworking the everyday fabric of urban life.

4.1   Assembling the Urban Periphery

As an assemblage of governmental technologies, the current approach to managing labour migration in South Korea through the EPS seeks to actively shape the presence, practices, attributes and tenure of migrants. As I have already noted this is a policy that is best characterised as 'migration without settlement' (Seol & Skrentny 2009: 582) because the emphasis is very clearly on labour market incorporation and the foreclosure of other possibilities for inclusion in society. As in other guest worker arrangements individuals arriving through the EPS are subject to a differential politics of inclusion and exclusion that generate 'friction' (Tsing 2005) in everyday life. Through border crossing migrants encounter processes of subjectification in which they are expected to participate only in certain predetermined segments of society and economy and where governmental technologies seek to foreclose possibilities for interpenetration (Yeoh 2006; Zhang, Lu & Yeoh 2015).

Despite the seeming proliferation of governmental technologies and the pervading influence and normalisation of political rationalities of migration, governmentality scholars often emphasise the importance of questioning the taken-for-granted character of governmental practice (Brady 2014; Dean 2009; Rose 1999). Arguably, the governmental desire to control and regulate migration is a response to the ongoing subversion of the nation state's borders either literally in the case of undocumented migrants or politically in the movements to recognise greater rights, challenge work relations or articulate alternative identities. As Mezzadra and Neilson (2013: 93) note in their account of multiplication in border technologies, it is critical to recognise that migration management is never a complete and coherent project but rather one that is always vulnerable to subversion:

> On the one hand,… multiplication plays the role of 'divide and rule'. On the other hand, living labor has still the chance to refuse to subordinate itself to the norm of abstract labor – or at least to negotiate its subordination. It is from this point of view that multiplication can become an incalculable element in the relations between capital and labor, giving rise to unforeseeable tensions, movements and struggles

It is exactly this subversion of power and control, or at least the negotiation of political subjectivity, that I am concerned with in this chapter. In order to address this issue, however, I seek to shift scales from the nation and 'migration management' as a governmental project to the level of everyday life, the ordinary and extraordinary presence and practices of migrants as urban residents. The chapter focuses on the differentiated way in which migrants become part of urban life and the manner that this links to the migration regimes that they move through. The emphasis here is particularly on migrant workers whose lives are shaped around marginal or peripheral spaces in the city but it is also imperative to note that the urban geographies of both teachers and students are also shaped by migration regimes, albeit with very discrepant outcomes. This varied urbanisation of migration involves the establishment of what Ong (2007: 85) captures in the term 'graduated sovereignty', 'the re-zoning of national territory into political spaces facilitates articulation with global capital, and also enables differential governing of groups and populations'. The claim that I advance in this chapter is that while political rationalities and governmental technologies constitute a subjectification of individuals crossing borders, not least through strategies of enclosure and the placement of migrants in central or peripheral spaces of society, these are never complete but are rather subject to negotiation or even contestation by migrants themselves. Because the focus in this chapter is primarily on EPS workers the chapter also builds on an understanding of the urban periphery as a geographical as well as social and political space in the city.

It is common in discussions in migration and urban studies to take the notion of the periphery literally, where it refers to nations at the edge of world or globalising systems, or to the literal outskirts of cities (Andrucki & Dickinson 2015). While such accounts of the periphery usefully highlight the positioning of different subjects in relation to the centre, they also serve to overemphasise the exclusion of the peripheral, and in the process undermine conceptions of agency or the possibilities that are also inherent in the periphery and peripheral subjectivities. Rather, the periphery, both as a geographical and socio-political terrain, also generates spaces for negotiating and at times destabilising these very projects of the state.

In taking this approach I build on Simone's (2010) conceptualisation of the urban periphery as a site of necessary tension, a potentially generative site of innovation and adaptation, but also simultaneously the target for strategies that seek to foreclose these possibilities. This tension emerges because the periphery often refers to a place that is part of the administrative domain of a nation, city or region but is not fully encompassed by that entity. In this respect, while the periphery falls under the jurisdiction of 'the centre' and is subject to its laws it is never fully brought under the regular logic and development interventions of the centre. For Simone (2010: 40), this position means that the periphery is also often a threat because it falls outside of the fundamental concerns of the centre but then is sometimes subject to excessive intervention:

> … these areas are sometimes in need of special development, humanitarian intervention, restructuring, investment, policing, or emergency controls. Thus, the periphery both is ignored and, at the same time, occasions excessive attention. It forces the state to admit to the necessity of exceeding its own 'core' values and technologies in order to rein it in, as perhaps the periphery's only concrete indication of its relevance

Critically, Simone also seeks out a conception of the periphery that draws attention to its possibility for assembling the city otherwise, for the innovative and adaptive work of urban residents in making their way in the urban worlds they inhabit and reworking the relations that marginalise them (see also Roy 2011).

Migration generally and the management of guest workers in particular represent a useful elaboration of this notion of the periphery. Indeed, as we have seen in the previous chapter migration undoubtedly falls within the scope of national regulation and the practices of the state but much of the state's interest is in managing migration in order to serve the needs of the populous who are not migrants. The general targeting of population, which as Foucault (1991) has informed us is the *raison d'être* of modern government, the enhancement of their wealth and wellbeing, is at best partially applied to migrant populations generally and only marginally to guest worker populations. Migrant populations are often ignored, certainly in terms of their daily lives and experience in society. Yet, at other moments, often associated with a crisis or problem in migration, migrant populations become subject to 'excessive attention' (Ghorashi 2010): they are incarcerated or deported (Gill 2016); targeted in settlement, integration or assimilation programmes (Schinkel & Van Houdt 2010); or become subjects of ongoing and insistent study and analysis as special members of society (Hess 2010).

Migration also comes to ground in particular sorts of ways that are necessarily geographically and historically specific. Guest worker regimes like the EPS seek to tightly constrain the everyday life and future aspirations of migrants. They seek, in other words, to generate 'docile subjects' who can be inserted and then removed from labour markets when necessary. While the desire for control operates principally through practices of bordering, the moulding of mobility and the eventual departure of 'temporary' migrant workers, it is also often articulated at the level of everyday life. Indeed, many guest worker regimes stipulate the locations for work, housing, time-off and even sometimes day off activities. In the Singapore context, for example, Abdullah (2005) argues that space plays a key role in the positioning of migrants in society. During the day, workers labour on construction sites away from interactions with the public, and at night they are housed in employer-arranged dormitories that are located on work sites or in industrial districts, or are closed off from public sight by gates and fences. For Abdullah (2005: 228) these arrangements construct a 'total institution': 'all phases of daily activities, like any total institution, are scheduled, timetabled and enforced, cumulatively brought together to realize the discursive ideal of the "good docile worker."' The contrast with international students inhabiting the spaces of the global campus (Sidhu et al. 2016) or English teachers residing in ordinary suburban spaces (Collins 2016) is considerable in terms of both the spatial and temporal characteristics of urban life. Nonetheless, what is clear across all these forms is that the everyday regulation of migrant bodies is an important element of establishing their differential position in the city where they can contribute to the productive work of businesses without disrupting the normative orientations of urban life (see also Gogia 2006; Elsheshtawy 2008; Collins 2012; Ye 2017).

Simone (2010) cautions us about demarcating a spatial location for the periphery but does note that it is the interstitial zone of the peri-urban interface that perhaps best articulates the conceptual and practical promise of the periphery. These zones of exception are certainly part of the city's overall administrative or political–economic domain but they are rarely fully encompassed in the political projects that seek to constantly remake the city. Where the urban core, financial districts, consumption spaces and new residential developments represent spaces of significant intervention in the crafting of global city spaces, the periphery often sits outside of the imaginings for urban futures. It is for this reason that 'activities and ways of living persist that are not completely integrated into city life' (Simone 2010: 39), that reflect wider urban processes but are not incorporated into normative urban systems.

> Residents find particular ways of dealing with those absences in particular combinations of generosity, ruthlessness, collaboration, competition, stillness, movement, flexibility, and defensiveness. Some residents will hold their limited ground to the end; others will live lives all over the place, willing to become anything for anyone almost anywhere. What these combinations will look like depends on the particular histories of cities and their relationship to other combinations – i.e. political culture and contestation, and how particular localities are situated in relationship to others, as well as other cities, regions, and economic poles. (Simone 2010: 24)

In other words, the periphery draws our attention to the manner in which residents but also migrants can never be fully disciplined or governmentalised subjects – their desires and everyday lives are always pregnant with other possibilities that exceed the control that is articulated on them. In order to tease out some of the ways in which the urban periphery is a space for the generation of a different politics of migration I focus here on the ways in which migrant lives both articulate with and also exceed the normative expectations placed upon them by the EPS and other migration regimes.

4.2   Migration and Marginalisation

In the South Korean context, the spaces within which migrants arriving through the EPS are typically situated are also very literally as well as ideologically generated in the urban periphery. The majority live and work in the outskirts of the Seoul Metropolitan Region in spaces that are an intersection between urbanisation and rural–agricultural practices, purpose built factory zones and minimally formalised residential developments. As Figure 4.1 illustrates, migrant workers are heavily concentrated in particular industrial areas of Gyeonggi province to the

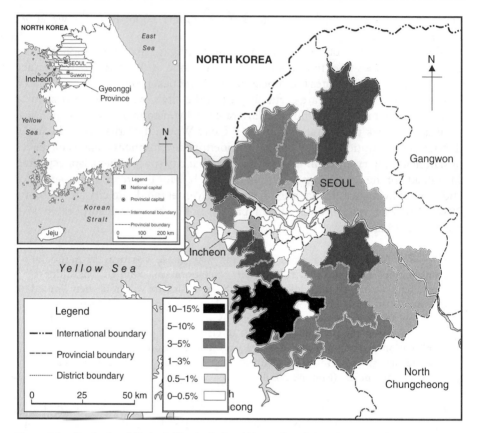

Figure 4.1   Distribution of E9 visa holders in the Seoul Metropolitan Region. Percentage figures indicate the proportion of the total E9 visa holders living in each area. Source: Korea Immigration Service. Map drawn by Lee Li Kheng.

north and south of Seoul city, and in neighbouring Incheon city. Although technically within the metropolitan region and sometimes within only a few kilometres of major residential, commercial and consumer districts many of the spaces that migrants work and live in are thoroughly disconnected from the centres of urban life in Seoul. Migrants invariably live on or near factories in makeshift accommodation assembled from light materials, containers or cut out of factory spaces; or in small family-run *gisuksa* (dormitories) that provide housing to meet employers' obligations. Without access to private vehicles it can take hours to reach even minor urban centres because public transport rarely travels near these areas and migrants have few other mobility options.

Although such areas form part of the broader metropolitan or capital region, they are characterised by quite different sorts of environments – much more

Figure 4.2   Workplace in Gyeonggi Province. Source: Author's photo.

sparsely populated, with areas of industrial development, agriculture and pockets of dense new residential developments. Unlike Seoul city proper the landscapes that many migrants encounter are often described as 'empty', 'quiet' and 'rural':

> At the very beginning, I didn't know anything. As in, nothing. Everything I knew, I learned through the pre-departure training. Like how there are four seasons. Actually, I expected Seoul to be a city. Well, of course it does have a city atmosphere, but I never thought that I would be going to a mountainous area. The factories are located in the mountains, you see. What I thought was that companies would be placed in neighbouring complexes, like my workplace in Cavite [in the Philippines]. That was my expectation. Which was why I was so surprised. The company that hired me was located at the isolated side of a mountain. So, it was lonely. (Bayani, Philippines, Male, EPS worker)

As Bayani's description here, and Phuoc's earlier account, suggest the lives of many migrants are characterised by a very real distance from the city itself and often also from daily encounters with other urban inhabitants (see Figure 4.2 for an example). Many of those in this research lived on site at workplaces and of those who didn't modes of transport included being collected by employers, walking or cycling. These environments are characterised by a paucity of activities: 'There are a number of small factories from one to three storeys in my area but it is not like an industrial estate. There are few houses and a grocery shop in the area.' (Nakharin, Thailand, Male, EPS worker). This distance tends to restrict the non-work mobility of migrants, making it difficult to travel away from workplaces even when there is time off: 'If there is no OT [overtime], I will leave from Friday evening. It takes around three hours by train. It will take longer if I go to a farther area like Gangnam district [a business and nightlife district in Seoul]' (Adul, Thailand, Male, EPS worker).

Peripherality is also constituted in the temporalisation of migrant lives in these spaces. While Korea is known for its very long work hours, the narratives of migrant workers highlighted a particular level of regimentation in the workplace within which the day started with early morning exercises, followed by extended hours of work late into the evening and considerable obligations to complete overtime.

> I arrive at the factory at 8 am and start my work from 8.30 am to a lunch break at 12 pm. I start my work again from 1 pm to 5.30 pm. I may need to work overtime until 9 pm. Then, I go back home. Although Sunday is supposed to be a day off, I may be required to work if there is extra work like today. I usually do not go out often. (Tidarat, Thailand, Female, EPS worker)

While the specific routines of workers varied considerably, including both day and night schedules, like Tidarat very few interviewees had time for non-work activities during the week and some regularly worked seven days. In many respects, this reflects the demands of relatively precarious small businesses that must operate on a just-in-time basis and often do not have the scope for lost productivity. For migrants, however, they also reinforce their position in the periphery because they reduce opportunities for social contact outside the workplace and with most accommodation on or near worksites almost all time is spent oriented to their role as workers rather than providing scope for incorporation as urban residents (see Figure 4.3).

Figure 4.3   Banweol industrial park, Ansan city, Gyeonggi Province. Source: Author's photo.

The urban geographies of migrants arriving through the EPS, oriented to the periphery, contrast considerably with those of English teachers and international students. For English teachers (Figure 4.4) the location of academies, schools and universities across the metropolitan region means that they also have a wide variety of urban experiences in their daily lives. There were only a few residential areas where there were even minimal concentrations of English teachers: Yongsan and Gangnam in Seoul, and Seongnam in Gyeonggi Province. These locations correspond with areas of affluence (in the case of Gangnam and Seongnam) and with neighbourhoods such as Itaewon (located in Yongsan) which has a substantial history of foreign settlement and amenities (Kim, E. 2004; Song 2013). English teachers are ordinarily provided housing as part of their employment contracts, or alternatively given an additional payment to cover housing. Generally

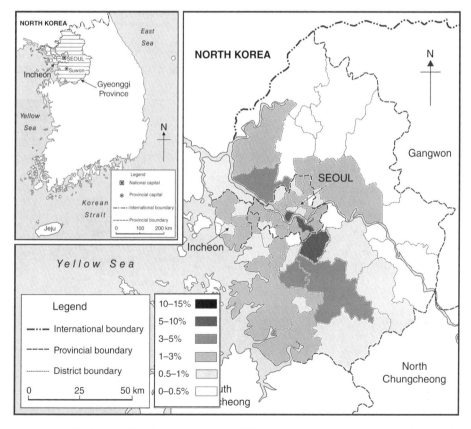

Figure 4.4   Distribution of E2 visa holders in the Seoul Metropolitan Region. Percentage figures indicate the proportion of the total E2 visa holders living in each area. Source: Korea Immigration Service. Map drawn by Lee Li Kheng.

they live in small apartments or in 'officetel' studios in complexes that can have a number of other sojourning residents living in them. While these arrangements are generally quite modest they are substantially better environments than those that EPS workers live in and are generally located near to public transport and other urban amenities, they are also typically situated within wider residential areas. Across this diversity of urban locations, the most typical neighbourhood experience is life in an ordinary suburb in the metropolitan region (see Figure 4.5 for example). Daniel (Canada, Male, English teacher), provides an indicative illustration:

> It's actually nice where I'm living now – it's a big … I would sort of almost sort of call it the suburbs of Seoul, you know a lot of apartment buildings – massive square kilometres of nothing but apartments but also very, a community – there's like a lot of community centres and a lot of shopping districts and nothing like COEX, … enough people but you know, it's quiet enough that there's not a lot of traffic and it's still nice.

While Daniel intentionally describes his experience of Seoul in contrast to urban spectacles like COEX (a massive convention centre, commercial and

Figure 4.5   An 'ordinary suburb' where English teachers live. Source: Author's photo.

shopping centre located in Gangnam), his account nonetheless draws attention to an indicative difference with the urban experiences of migrant workers. Except in a small number of instances English teachers in this research almost always lived in neighbourhoods alongside Koreans rather than in distinctly foreign spaces or within their workplaces. This location presents opportunities for encounter that are more ordinary and that also open possibilities for building relationships with local residents and developing attachments to place.

The urban geographies of international students were unsurprisingly oriented around the universities that they studied at. Figure 4.6 presents the pattern of D2 visa holders in the Seoul Metropolitan Region. Reflecting the location of the most highly ranked universities (including KU and SNU), international students are largely concentrated within a small number of districts within Seoul city. The

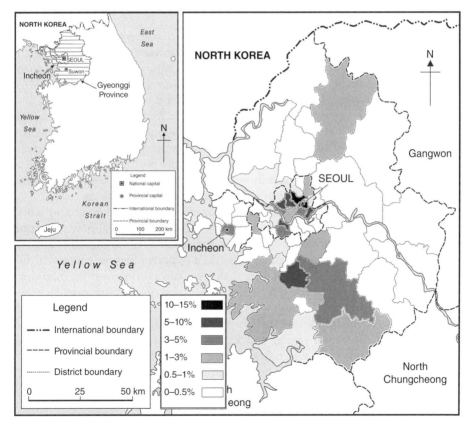

Figure 4.6   Distribution of D2 visa holders in the Seoul Metropolitan Region. Percentage figures indicate the proportion of the total D2 visa holders living in each area. Source: Korea Immigration Service. Map drawn by Lee Li Kheng.

central districts of Dongdaemun, Seongbuk and Jongno, which together accounted for 8,428 international students in 2012 (around 10% of all students in South Korea), host a number of major universities including KU as well as Sungkyungwan University, Kyung Hee University, University of Seoul and Dongguk University. This grouping of central districts includes KU in Seongbuk, which is the location of the second largest number of D2 student visa holders in Seoul at 2,939. Gwanak (south of the Han River), the location of SNU, had a smaller number of D2 visa holders at 1,327 (making it 7th in Seoul) in 2012. This partly reflects the absence of other major universities in that area but also the somewhat more isolated character of SNU that leads some students to live outside of Gwanak district. Outside of Seoul City itself there are also smaller concentrations of students in areas where major universities are located: Suwon and Yongin in Gyeonggi Province (both of which host local universities and satellite campuses of major institutions from Seoul) and Namgu in Incheon (location of Inha University).

Many of the areas where international students are concentrated represent key sites in the crafting of Seoul as a global city, where universities play an important role in putting the city on the map and creating attractive urban areas for education, culture and social activities (Goddard & Vallance 2013; Sidhu et al. 2016). The adjoining districts of Dongdaemun, Seongbuk and Jongno, for example, host a number of major universities and are also the site for several landmark urban developments: the Cheonggyecheon stream, Dongdaemun Design Plaza, Daehangro theatre district and the Insadong and Samchang-dong heritage areas. In this respect, international students are very literally positioned in the centre of Seoul's global city aspirations, and in many respects embody a key feature of the desire to encourage the presence of certain kinds of mobile subjects.

Set alongside each other these three examples of the residential and work/ study patterns of workers, teachers and students in Seoul reveal how migration regimes intersect with urban geography to shape the lives of migrants, both in terms of where they are positioned and also their opportunities for encounter that are discussed later in this book. Rather than a backdrop or context for migratory processes, the urban landscape is an active constituent of the differential incorporation of mobile subjects. There is a regulatory gradient to these differences, with migrant workers inhabiting the urban periphery, English teachers living and working across the urban region and international students positioned in the urban core. The state plays a significant role here in setting the conditions of different visa types and the relations between migrant workers and English teachers and their employers, and to a lesser degree between students and universities. At the same time these spatialities also demonstrate broader political–economic transformations in Seoul, where most manufacturing industries have increasingly shifted to the outskirts of the region, English language academies have become a normal part of residential areas and the most internationally oriented universities have emerged in the centre of Seoul. These different urban spaces clearly offer varying opportunities for mobile subjects to participate in urban life – either in terms of being visible in public spaces, interacting with local populations and other migrants or contributing to longer-term changes in the spaces of the city. For those who live in the periphery who are the main focus of this chapter, it is social networks and practices of care that matter in generating possibilities for reworking what is possible in the city.

4.3   Generating a 'Mobile Commons' in the Periphery

The encounter with the spatial and temporal dimensions of the urban periphery generates friction in the lives of many migrants arriving through the EPS regime, circumscribing their daily activities to particular spaces and regimented activities and reducing the prospects for wider practices of inhabitation. The urban periphery and other spaces in the city are not only constituted in advance of migrant arrival but also through shared experiences with other migrant and non-migrant populations who are immersed in these worlds. The sociality generated in encounters with others is a critical resource for migrants seeking to renegotiate their position in urban life and to craft new aspirations and opportunities for themselves. Connections with others, particularly other migrants, can allow individuals to be mobile when they want to be and make it possible to establish more conducive living and working arrangements. Social connections, then, are a resource or 'infrastructure' (Simone 2004; Xiang & Lindquist 2014) for getting by in daily life and reworking peripherality. The examples of Thuy (Vietnam, Female, EPS worker) and Pin (Vietnam, Male, EPS worker) offer useful illustrations.

Thuy had been in South Korea for seven years at the time of the interview. Her migration was generated through the brokerage of a 'fake marriage' to a Korean man and her own sense of the opportunities that this would make possible, 'because I wanted to work to support my family [in Vietnam]'. The marriage lasted six months: Thuy had frequent disagreements with her husband and was subjected to domestic violence. Fleeing her marriage in Busan and without legal status outside this relationship she sought safety in the social networks of Vietnamese in parts of the Gyeonggi province and then subsequently through relatives. She did not wish to travel back to Vietnam because she knew it was unlikely she would be permitted to return. Having left Busan she moved into an apartment with seven Vietnamese women who were in a range of situations – some documented and undocumented workers, others like Thuy having left marriages.

> I lived for one year here just staying at home without working because I did not dare to collect my documents at that time [due to fear of deportation] and employers wouldn't hire me without legal documents. If I was lucky, they might hire me but

just for a few days, my earnings were so little that I had to borrow money from my friends. Luckily, I found my aunt; she guided and supported me a lot in living here.

Thuy's narrative spoke to the gendered division of migration into South Korea. Migrants from Vietnam, for example, include those who come through the EPS regime, who are overwhelmingly men, as well as women like Thuy who migrate into brokered marriages with Korean men but are also often oriented towards work and livelihood (Kim, M. 2013). As Thuy's account shows us this gendering of migration is very significant for the social lives of migrants and the rights that they are accorded. She also highlights the significance of social relations and shared resources in the negotiation of friction in migration, in avoiding this violent relationship and establishing a different set of opportunities as an undocumented but perhaps freer migrant in Seoul. Thuy's friends and aunt helped her to secure a job with a car accessories factory where the employer accepted undocumented workers. While the work was difficult it has provided opportunities to adapt to life in Seoul and to establish alternative possibilities for remaining in South Korea.

Pin had been in South Korea for eleven years at the time of his interview. He is an undocumented migrant and as the next section makes clear an individual who has been very successful at reworking his peripheral position as a migrant worker. His narrative also speaks to the significance of social relations and shared resources amongst migrants for the negotiation of the complex landscapes of migration. Pin didn't know much about Korea but when he had looked for opportunities to work abroad had been offered opportunities in Libya or South Korea; he recalled that 'Libya sounds strange and I heard that the weather there is so severe, I felt scared so I chose to go to Korea'. He was initially contracted to an employer in Pohang, a centre of heavy industry in the Southwestern part of the Korean peninsula, and worked in a factory there for six months. The wages were low and the conditions poor and like many workers under the ITTP Pin soon abandoned his contract and sought new opportunities in and around Seoul. He recounted how he was advised to do this but also how he was provided information on how to find a job, accommodation and to avoid authorities during immigration crackdowns:

> It was a precarious place in the company. I received support from some of my friends who moved out before, actually there were friends who could help, and sometimes I myself had to pay for the information. For example, they helped me to find a job then I had to pay commission for them.

These social connections and the information that is generated in and circulates through them are a critical feature of migrant life generally but in particular for undocumented workers who cannot utilise official sources. As Pin's excerpt reminds us, these relationships, while potentially beneficial for migrants seeking to direct their own mobility, are not always offered altruistically, but rather also rely on economies of exchange, often monetary. Whether offered freely or as part of an exchange such information rests on accumulated knowledge of how to negotiate migrant life in Seoul and South Korea, knowledge that relates to personal experiences of individual migrants but also information that has been circulated through social networks of migrants who have come and left South Korea over many years. Papadopoulos and Tsianos (2013: 190) describe these politics as the 'mobile commons':

> People on the move create a world of knowledge, of information, of tricks for survival, of mutual care, of social relations, of services exchange, of solidarity and sociability that can be shared, used and where people contribute to sustain and expand it.

These are social relations that facilitate but also mediate mobility, that allow migrants to negotiate the complex landscape of regulatory regimes, the unequal organisation of workplaces, and the social, cultural and economic means of survival while on the move.

The mobile commons then can be understood as a key infrastructure that is involved in linking forms of desiring-migration to shifting opportunities in the city. Perhaps most notably for the purposes of this chapter is the manner in which this commons is constituted through circulations of information about the different lives and conditions of other migrant workers. Different wage levels, employment practices, rules and regulations, self-help groups, accommodation standards, places to eat and buy food, religious worship are all discussed on a regular basis amongst workers. Such information opens up possibilities for sustaining migration or altering the ways in which migrants inhabit urban spaces towards more desirable modes of life. Often these connections are formed through social networks that work through national, ethnic or even village identities and associations that form within specific workplaces and neighbourhoods. At other times, social networks are formed across wider territories and incorporate a diversity of migrant workers or even other foreigners in South Korea or Koreans concerned with the wellbeing of migrants. Nonoy (Philippines, Male, EPS worker) provides one example from Ulsan before he moved to Seoul:

> So, there wasn't any organisation there before, because there weren't a lot of Filipinos, so I was in Busan, with a Filipino migrant workers association. Then we formed we tried to help Filipino women with Korean husbands. And we formed what we call USAP or the Ulsan Association of Filipinos we were the ones who first established USAP. Until now, it is existing since we were the, our purpose [was to create] a community of Filipinos.

Later on Nonoy was asked to take a role in a publication that is written for Filipino EPS workers and other migrants in South Korea, an opportunity that saw him take an increasingly central role in community activities. His increasing involvement in these forms of the mobile commons reflects an intensification of his investment in migrant lives and networks in South Korea, a reorientation of desire from the possibilities and excitement of migration towards becoming otherwise: an advisor and community leader.

Like Nonoy's example of becoming part of organisations that support the lives of migrants these efforts, even as they are tied to institutions like the university, reveal the ways in which different migrants recognise connections with others who are similarly seen as foreigners and seeing value in the building of solidarity and mutual recognition. Many participants articulated a view that 'we are foreigners', 'that is the life of foreigners' and that 'we have the same conditions'. Many also spoke about looking out for others, trying to help new arrivals and standing up collectively to employers who were abusive to fellow workers. In the most hazardous of situations, it is this 'politics of care' (Piper 2006) that motivates migrants to put themselves at risk for others, to support the evasion and invisibility of fellow migrants:

> I felt so sorry for those five men. They were treated badly by their *sajang* [manager], such as being forced to do something, kicked, beaten, anything. The *sajang* even gave an order by hitting them using a wrench, in the cold winter. So what did they want? Of course they wanted to escape from that company. So, I helped them to plan the escape strategy. They all came from the same company in Indonesia before. I met their agent once, so I wrote him an apology for helping them escape from their company in Korea. ... At first, I took some of them to [neighbourhoods in Gyeonggi], and others to [neighbourhoods in Incheon]. That's all I could think of. (Adit, Indonesia, Male, EPS worker)

While examples of solidarity and social network formation were more pronounced in their significance amongst EPS workers there were also examples that emerged in the accounts of international students and English teachers. Amongst international students, for example, many participants spoke about nationality-based student associations, usually informal, that helped newly arrived students learn about Seoul and their university, provided advice and mentoring and organised social activities for students. There are also organisations like the Korea International Students Association (KISA) and SNU International Students Association (SISA) that link students across multiple nationalities and that at times reach out to build wider networks with other migrant communities. These commonly take on institutional support from the university and are channelled towards providing information about campus and city life, running cultural festivals and other events for students.

Another example of the mobile commons can be distilled from English teachers in a way that adds to our understanding of how such practices matter in articulating migration to urban life and reconfiguring the city to the desires of migrants. Allan (Canada, Male, English teacher) recounted the injury and death of an American teacher Bill Kapoun in an apartment fire in 2008 and the subsequent response and collective action.[1] Friends of Kapoun set up a bank account for donations to support his hospital bill that he was liable for because his employer had not registered him for health insurance. While Kapoun eventually died and his parents started a fund with the remaining money, the response eventually led to the establishment of the Association of Teachers of English in Korea (ATEK), described as a voice to represent teachers:

> Strength in numbers…when there is a huge change in the way immigration functions. When the government wants to go to foreigners and ask what do you think about this we hope they go to us. That doesn't mean that we are going to be beating down their door everyday but we at least want to say that we represent X number of members, let's say 1000 members, we hope more, right now around the 600 mark, so we could at least be a voice for those people and they can be confident someone is at least asking the right questions. (Greg Dolezal, former ATEK president)

An organisation like ATEK works by drawing attention to the needs of a particular group that are not recognised and for providing 'a voice for teachers' (Tom Rainey Smith, founding ATEK president). It did not take the form of protest action or a workers' union, but rather a kind of lobby group that represented English teachers. During the three years that ATEK operated the organisation established shared resources for newly arriving teachers, engaged in an 'equal checks for all' campaign to end HIV and drug testing for English teachers[2] and provided avenues for individuals to respond to discrepancies in employment practice. In this respect, ATEK also sought to reconstitute teachers as not only sojourning travellers but also workers with rights and individuals who could make substantive claims about their place in society. The relatively short life of ATEK (dissolved in 2012 after key members withdrew and left South Korea) also says something about the transience and churn of English teachers, that will be discussed in greater detail in Chapter 6 and that has significant bearing on the capacity to generate a mobile commons. Moreover, in comparison to accounts of migrant politics discussed in Section 4.4, ATEK was also limited in its duration and impact exactly because teachers do not experience equivalent marginalisation as EPS workers, nor is the same imperative there to become visible through interventions in public space.

We can productively understand these social relations as an example of what Simone (2004) has elsewhere described as 'people as infrastructure', the collaborative activities of residents who otherwise seem marginalised from and disenfranchised by urban life. Social networks and actions of solidarity are a technique for addressing marginalisation in day-to-day life, as is a common situation for EPS workers, or to overcome unexpected disruptions to the more stable lives of other migrants like English teachers. In both cases this social infrastructure directly addresses gaps or disruptions in the more stable or physical infrastructure of urban life, workplaces and accommodation. Of course, as I have already noted, these relations are not only built on a sense of care, but also at other times efforts to profit from relations, to earn money through the provision of information or connections that might advance one person's life. The knowledge and resources that are circulated through such channels are accumulated over long periods of time, they have been built up through years of migrants coming to South Korea for varying periods of time, learning from experiences, adapting, finding responses and then sharing that information on through further connections. They represent, in other words, a key social component of the ways in which migrants become part of the assemblage of urban life, linking transnational migration trajectories and associations to daily life in the context of differential inclusion through the migration regime.

4.4   Becoming Undocumented and the Subversion of Control

One of the central objectives of migration regimes like the EPS is the eradication or removal of migrant subjects defined as 'illegal', 'undocumented' or 'irregular'. The desire to regularise migratory processes does not, however, simply respond to predetermined categories but rather is in itself constitutive of the lines between who is legitimate and illegitimate within the space of the nation (Kalm 2010). The concern for managing migration, then, also involves active practices of legalisation and illegalisation. Yet, in spite of the increasing efforts of the South Korean and other governments to regulate the space of the border, and to manage the presence and practices of migrants within the space of the nation, the migration and desires of migrants are rarely captured by even the most intricate of migration regimes (Dauvergne 2008). Rather, unauthorised migration is pervasive in tactics to avoid detection in border crossings, to remain in place beyond time-limited visas and permits, and to engage in social, economic and political activities that are presumed to exist beyond the realms of migrant lives (Seol 2012).

In the case of this research, accounts of becoming and negotiating undocumented status were only present amongst EPS workers, but several English teachers referred to 'cowboy teachers' who worked on tourist visas and there were some international students who had worked in excess of the hours permitted in their visas. It is migrant workers, both authorised and unauthorised, who because of their marginal position must actively negotiate migration regulation through strategies to evade detection, acceptance of rules for short periods of time and the accumulation and circulation of knowledge about how state apparatuses and wider governmental arrangements operate. Jaya (Indonesia, Male, EPS worker) has been working in South Korea on three occasions since the mid-1990s, moving between being documented and undocumented on numerous occasions. Arriving initially on a Trainee Visa, like Pin, Jaya was soon informed by other migrants that much more money could be earned and freedom secured by becoming

MIGRATION, THE URBAN PERIPHERY AND THE POLITICS OF MIGRANT LIVES 91

undocumented because he would then have scope to negotiate wages and leave employers if they treated him poorly. In 1998, he left South Korea after staying over the time limit and breaching conditions of his visa. When he decided to return in 2001, Jaya arranged for an alternative passport with a different name, Wijaya (he literally added a common prefix to his name), and entered on a tourist visa with his wife and quickly returned to work. It was during this time that the debates around the EPS were becoming stronger and before it was implemented there was a crackdown on undocumented migrants and an amnesty visa for those who agreed to depart within one year. Jaya took this amnesty visa and then remained for another three years in South Korea until 2005, departing again with breached visa conditions. In 2008 he returned through the official EPS system, using his real name and remaining documented until the time of this interview:

> The most important thing is the visa first. My visa expired at the end of 2004, and at that time we can get recommendation letter [from the Ministry of Labor]. So it's illegal, then legal again, illegal. When I wanted to go home I got the recommendation letter. But because I decided that I would not go ahead and study [in Indonesia], I ignored that letter. I added several months because it had no use, really. That recommendation letter was useless. ... After that they knew that I was illegal.

Jaya's narrative speaks to the capacity for migrants to direct their own mobility in ways that exceeds that which is predetermined by the state and other governmental actors. Rather than leading to the eradication of unauthorised migration, strategies of exclusion and exception engaged in by the Korean state in the early 2000s generated both illegalisation of thousands of workers but also the active undermining of these strategies by individuals like Jaya (Seol 2012).

The peripheralisation of labour migration and the lives of migrants in Seoul is a critical enabling feature of the sorts of tactics employed by migrants like Jaya to remain in place, to earn more money or to engage in freedom of movement. As the description of the urban periphery in Section 4.2 suggests, the location of many migrant workplaces and residences are a considerable distance from the centre of Seoul or even from significant built up areas. There are often only other factories in the immediate vicinity, migrant housing is often within or next to factories, and there are few opportunities to encounter Korean residents not involved in these workplaces. There is, in this respect, an alignment between the figurative invisibility of migrants who have become undocumented, they are no longer visible to the surveillance of the state, and their very literal efforts to be invisible in their everyday lives. The Filipino acronym TNT ('tago ng tago') literally means 'always hiding' and is an informal term for illegal migrants that encapsulates the importance of invisibility in the urban lives of unauthorised migrants. This is not a situation to romanticise. Indeed, undocumented migrants or their colleagues often spoke about the need to remain only at work or in their workplace housing, to negotiate access to different places through reliance on friends, relatives or strangers who simply happen to also be migrants, and the constant feeling of vulnerability, of the risk of being exposed and the potential consequences for their future status.

The presence and practices of unauthorised migrants in Seoul's urban periphery also relies on and generates alternative social relations of production that can both disempower and empower these migrants. On the one hand, becoming undocumented exposes migrants to potentially exploitative employment practices where they have no recourse to the law and their very invisibility in the urban landscape means that this exploitation is rarely exposed. Beyond the surveillance of the EPS regime and the labour protections that have been embedded in it, employers can choose to leave wages unpaid, ask for extra work without extra pay, provide substandard accommodation or even engage in verbal or physical abuse. Nugi (Indonesia, Male, EPS worker), explains the uneven contours of what he calls 'private work':

> If we do the job well, we will continue working. But if we don't, that would depend on the factory's decision. [Do you return straight home?] No, we don't. We try to find another job.... Oh, usually it was about language problems. I couldn't explain about something. Once I was hit with t-shirt threads. It was because I put the wrong threads, the colour was similar but actually different. I could have reported that. But it was just a waste of time. It was not painful anyway. Well, that's the issue. The problem of private working, sometimes the *sajangs* [managers] do whatever they like. Sometimes we are not paid as we are supposed to. The sole authority was in the hands of the *sajang*.

Such accounts are not surprising on the one hand and fit neatly within the narratives that the architects of the EPS have created about the need for a transparent landscape of migration where migrants are paid a minimum wage, have clear rights and a recourse to the law in cases of abuse (Min 2011). And yet, not all unauthorised migrants experience similar circumstances. Others characterise their undocumented status as empowering, providing the capacity to negotiate with employers, to be paid more, and to have freedom to move without permission from the state. Bayani (Philippines, Male, EPS worker), for example, noted that there were many smaller businesses who have come to rely on undocumented workers and while some will also rely on exploitation others develop more reciprocal relationships:

> The big companies, the ones with big buildings, are not usually located in the countryside. So, when we say 'Seoul alone', we mean this place, and the treatment here is mostly okay. Because it's mostly small businesses. Even undocumented workers get support from employers. They're cared for. Because the companies here are just small businesses, Korean nationals seldom work here. So mostly it's small companies with just a few Koreans, then the rest are foreigners. Especially since the Koreans are picky about their jobs.

Some participants who were documented described how undocumented workers seemed to have more advantages than their documented colleagues:

It's as if they are the ones who are legal. They have their own rooms. …When it comes to salary, sometimes they give us half, but they give it in full for TNTs. (Sabrina, Philippines, Female, EPS worker)

This seemingly contradictory situation reflects in part the recognised risks for employers, who provide advantageous conditions to undocumented workers so that they are not exposed. In some cases, this is because the undocumented workers have established important roles in their workplaces, because they have had longer experience in Seoul or South Korea, have higher language proficiency or have acquired important skills.

Pin (Vietnam, Male, EPS worker), who earlier discussed his departure from a 'precarious' employment situation in Pohang, went on to find long-term and very rewarding work in Seoul as an undocumented worker. Through social networks and paid information he eventually found work in Seoul and over the course of the next decade established himself as a key employee in a garment factory where he now manages other workers including Koreans:

I am a foreigner and live illegally but I don't feel I am under threat, or deprived of the human rights. In general, I feel so comfortable. …To be honest, if the boss is not good, I won't work. I work there for years and this position is important. So not only the boss but all the people in the factory have to respect me. I work there for seven to eight years. My boss has not ever scolded at me, never ever. In general my relationship with the boss is good. We get along well at work as well. For example, we can work without talking to each other because we get along so well. In general, the relationship is good and they treat me well. … My relationship with my Korean [colleagues] is good [too]. Because, although I am a foreigner, actually I am their manager. I manage them, the Koreans do not manage me. So they respect me.

This is an articulation of considerable agency by a mobile subject, the undocumented worker, who is so often characterised as a 'victim', vulnerable and in need of assistance (Yea 2015). Migrants like Pin choose to remain undocumented in order to extend their time in South Korea because it is a good opportunity to earn further income but also because they have established substantive relations with employers, colleagues and friends. They have constituted a position for themselves, even if it is in the urban periphery.

Becoming undocumented in migration is clearly about the subversion of control as migrants are very literally undermining governmental attempts to manage the directions, durations and outcomes of migration. On the one hand, this is because becoming undocumented denies the state the information that is needed in order to manage migrants. Yet, at the same time, becoming undocumented also undermines the liberal imperative to 'protect' migrants, to set minimum standards for pay, work, accommodation and rights. In becoming undocumented migrants subvert the normative logics that presume they will be willing to inhabit a governmentalised position as an ideal mobile subject who works for appropriate pay and leaves at an appropriate time. Instead, unauthorised migrants expose themselves to risks and exploitation but often do so as part of a desire to control their mobilities, to constitute their own position in relation to the different spaces and possibilities of the city. The prevalence of undocumented migrants amongst EPS workers compared to teachers and students, then, is revealing of the extent of their peripheralisation and control by migration regimes, but also demonstrates how migrants are able to exceed these controls exactly because of their particular position in the city and nation. This is not necessarily a comfortable position but it can be one that empowers migrants by presenting them as, for example, productive workers or even friends even when they are expected to remain time-limited governmentalised subjects. Clearly the spatialities of the periphery are critical to these practices of subversion. They provide the invisibility that is needed to remain below the radar of surveillance both of the state but also of a range of others who might find the presence of undocumented workers troubling: residents, NGO and community groups who seek to protect migrants and even other migrants. The periphery's remoteness, few inhabitants and position outside of the visions of other urban political projects is key to generating and maintaining these invisibilities and making space for these alternative livelihoods.

4.5   Tactics of Recognition

For the most part the lives of EPS migrants that have been discussed hinge on invisibility, on migrants getting by in their mobile lives without drawing attention to themselves in ways that might problematise their place in the world. This has long been a tactic of migrant communities, to remain below the radar, either as a 'model minority' or by simply avoiding detection in border crossings, workplaces or the spaces of the city. In the context of a surveillant state that seeks to control and manage migration through the EPS and other regimes, detection can problematise the very mobility that migrants seek. At the same time, as Mitchell (1996: 61) has also shown, these examples indicate that invisibility can be 'both oppressive and potentially oppositional', particularly where it allows migrants to avoid or subvert state and employer attempts to rationalise mobility (Dauvergne 2008). In a similar way, visibility can also serve a dual purpose. Visibility can be seen as an important tool in the policing gaze of migration management, in the ability of state agencies and employers to place individuals in particular spaces and roles and to determine their involvement in society (Tazzioli & Walters 2016). At the same time, tactics for generating visibility by migrants themselves can also serve as a potential avenue for responding to the forms of domination that characterise labour migration regimes, particularly when such visibility challenges the normative placement of migrants in the periphery or other marginal urban spaces.

The lives of migrant workers in particular revealed examples of the ways in which visibility forms part of strategic actions that focus explicitly on challenging the differential inclusion of migrants. These actions include advocating for migrant rights in workplaces or through civil society or state institutions, developing migrant advocacy organisations or at times in public protests that challenge normative politics around migration and the rights of migrants (Kim, D. 2011). As scholars exploring migrant politics in other parts of Asia note these politics are not usually widespread amongst migrants (Ford & Piper 2007), not least because of the limited formal rights of citizenship that migrants are accorded. Likewise, in this research EPS workers commonly reported cases of exploitation but rarely described instances of challenging employers, seeking redress or engaging in protest. When they did occur, these actions included involvement in nationality-based organisations that advocate and support migrants (Nonoy, Philippines, Male, EPS worker); joining workplace protests to improve conditions (Thaklaew, Thailand, Male, EPS worker); and as the two detailed examples below indicate, acting as a mediator between Korean NGOs, the embassy and migrants in need (Adit, Indonesia, Male, EPS worker) and organising public protests (Monty, Philippines, Transgender, EPS worker). The relatively uncommon nature of these actions does not undermine their significance in challenging oppression and triggering important changes in the treatment of migrants (Constable 2009). Indeed, in South Korea it was migrant and civil society led protests that increased pressure on the government to develop the EPS and close-down the more exploitative trainee system that preceded it (Kim, D. 2011). Through advocacy and protest, migrants can achieve forms of visibility that they are ordinarily denied through their position in the periphery and their discursive framing as deviant, criminals or simply oppressed victims.

In order to elaborate on the politics that can emerge amongst EPS workers, I detail two cases of seeking to make migrant worker rights visible, in relation to state institutions (Adit, Indonesia, Male, EPS worker) and in public protest (Monty, Philippines, Transgender, EPS worker). Adit originally came to Seoul through the ITTP in 2001 and worked as a trainee until he returned to Indonesia in 2005, after which he applied through the then established EPS and came back to Seoul in 2006. Adit was well known amongst Indonesian workers in this research, not least because as he had 'been taking care of migrant workers matters since 2001'. His approach was to utilise relationships with the Indonesian embassy that he had cultivated over several years and to act as a facilitator of greater communication and to encourage advocacy on behalf of workers:

> I usually go there in a group, as I am focused on public relations. When the embassy becomes difficult I told them I bet there will be a demonstration soon. [Embassy

staff] asked me to keep any information about them, to allow them to be prepared, but I told him to gather embassy members and Indonesian migrant workers, to have a conversation regarding this matter. Otherwise, its prominent figures would rebel.... [Later] I joined the embassy committee. That's when I started to learn about passport activation, salary, and relationship with the embassy. The embassy has been giving a better service since then.

As he described it elsewhere in the interview, Adit saw his role as focused on 'channelling' the problems faced by migrant workers and the potential assistance that can be offered by the embassy. His actions then demanded that the embassy recognise the situation of migrant workers from Indonesia and establish systems to support them. It also led more recently to an approach from a Korean NGO focused on migrant worker issues who was seeking assistance to directly address issues of underpayment and workplace abuse, enabling a further set of connections: 'It opened my way to meet the Koreans, let's say their name are A and B. We make a good relationship until now. So I help them to take care of migrant workers' problems.' Tactics of recognition then are also about establishing alternative relationships in the city, with the representatives of nation states or with community organisations who can play a role in improving conditions.

Monty's experience offers another example that also speaks to the development of a migrant subjectivity as well as a politics that actively constitutes the migrant as a subject with claims to legitimacy *as a migrant* rather than an unformed or unofficial citizen. Monty originally came to Korea in 2006 and worked in a range of different factories. Like many other migrants Monty experienced sometimes repressive conditions both in the workplace and in the vulnerable position migrant workers have in the periphery. Shortly after arrival, Monty assisted a young Filipino co-worker to escape from a threat of rape by a manager in worker housing and helped to negotiate an early release from her contract. In 2009 in response to abusive employment practices in a computer factory, Monty organised a reluctant group of Filipino workers to stand up to their employer collectively with the threat of a lawsuit and demand changes to the way the workplace operated, the provision of overtime pay and the employer–employee relations. While initially facing challenges in getting agreement from nervous workmates the result was a rearrangement of the workplace and recognition of rights:

No one signed it. I was the only one who faced management. But I still went through with it. Because I still filed a case against the company. So the – because at the time I was studying – at the time we were already getting training [from the Migrant Trade Union (MTU)]. We had monthly education sessions on labour rights and other things. So during that time, I learned how to compute the overtime, I learned how to compute wages, I learned how to – I learned everything, like everything there was to learn. I even listed discrimination in the petition. And unequal treatment. The forced overtime. Because the way it was in the company was that you had to – go there to prove that you were really sick, before you could take a sick leave.

...So I listed them all down [our demands]. I won in the end. They did compromise. They asked me what I wanted to happen.... So that's what I said. So – at the time I said [to management], 'OK, I'll retract this but first, pay up the overtime, and second, make the rotation fair. So that – there'll be those to take the morning and the evening, but the rotation should be fixed so that – everyone earns equally. And then – then the discrimination from the manager,' I said. 'If you don't remove her, the discrimination will never end.' That's what I told them. The manager wasn't removed but she really did stop swearing until she eventually did resign.

Recognition here is achieved for migrants as workers deserving of fair treatment. As Monty's as well as Adit's statements suggest, these important changes were achieved through knowledge of labour rights and the systems involved in calculating wages and work time, information that was opaque to many migrants in this research. Organisations like the MTU are clearly also significant, providing opportunities to learn the skills necessary to challenge cases of outright exploitation. The MTU's aim is not only to be a voice for migrant workers in South Korea, including undocumented workers, but also a resource for self-empowerment (Kim, J.K. 2015). Enabled by knowledge of what can be claimed, the tactics undertaken by Monty and others make the importance of migrants in the workplace visible. The actions involve defining what is possible and challenging the authority of employers to set the rules and to use the skills circulated by groups like the MTU to enhance migrant awareness of their own rights.

Monty departed South Korea not long after this incident following visa expiry. After two months Monty returned to Seoul with a new interest in working but also in contributing to the community of migrant workers in South Korea. Monty became more involved in the MTU, eventually becoming a senior figure and led a number of movements to advocate for better recognition of migrant workers contribution, to the legalisation of the MTU as an institution and to improved workplace conditions.[3] As in many other stories of migrant activism in South Korea over the last two decades, visibility in urban space was critical to these protest movements – shifting the bodies of migrant workers from peripheral locations at the edge of the metropolis to the centre of the city where they cannot be invisible (Doucette 2013). In July 2010 as part of broader protest movements against the G20, the MTU established a sit in at Hyang-rin Church in the central shopping district of Myeongdong. The sit-in lasted over a month and included a hunger strike that drew significant media and political attention:

So August, from July to August, I lasted up to 30 days. I went through a hunger strike. So I went on hunger strike for a month. One month where – one month of just drinking water. Then afterward, the issue on the crackdown of migrant workers got bigger, because – there was a Vietnamese [woman] who was murdered around the same time, that was the time that a – a Vietnamese wife, she married a Korean, and within eight days, she was murdered by – what, by her husband – who was

Korean. That was part of it. *So it was like that was the time when – It got bigger and more visible, this issue.* About migrant – migrant workers issue, and also the migrant – the ones in general, the foreign workers, the foreigners.

The public protest described by Monty follows a pattern that has become common in key moments of migrant resistance in South Korea. Churches like Hyang-rin as well as the Catholic cathedral in Myeongdong have been used as sites for drawing attention to migrant exploitation and for pressuring governments and employers to change their practices. In addition to the support church groups have provided, these locations are significant because of the way they symbolically and materially rearticulate the struggle for democratisation in South Korea (Kim, J.K. 2015). Churches, particularly those in Myeongdong, were key sites for Korean democracy movements resisting authoritarian governments in the 1980s. Likewise, many migrant protests like the one described by Monty have tactically employed the earlier styles of the democracy movement by undertaking hunger strikes and displaying portraits of deceased and deported migrants in a commemoration of earlier sacrifices (Doucette 2013). As Monty indicates these protest tactics make migrant workers and their struggles 'visible' to the public. They force recognition of the ordinarily peripheral lives of migrants through the occupation and reappropriation of spaces that are geographically central in Seoul and historically central in South Korea. In this respect these actions seek to challenge the institutional framing of migrant workers' presence as abnormal and marginal, and in the process undermine the invisibility of migrant lives, challenging their peripherality by drawing attention to their centrality in contemporary Korea.

4.6   Conclusion: Urban Politics of Migration

The politics *of* migration as they have been articulated here are necessarily emergent rather than established, fixed and made normative. Whether in processes of unauthorised migration, the generation and maintenance of the mobile commons and tactics that demand recognition for migrant lives, these politics always and necessarily exceed what is expected in the governmentalities of migration management. Rather than becoming good docile subjects, whose rights are minimally protected but who will remain in place for predetermined periods of time engaging in prescribed activities, mobility and the desires it manifests often resist, transcend or escape the imagined futures of governmental programs. The task of governing mobility is complex and incomplete. Migration can be moulded by states, employers, communities, families and migrants, but their effects remain indeterminate.

The urban and more-than-urban settings of migrant lives are a critical terrain for the unfolding of these tensions between governmental desires for control and migrant desires for migration that suit their own ends. Indeed, the urban geographies of migrant lives discussed here reveal the discrepancy that is generated between migrant workers, English teachers and international students in the migration regime. These migrants' differential inclusion in the periphery, suburbs and centre of Seoul demonstrate how the social and legal construction of migrants as having different roles in economy and society comes to shape the manner that they become part of the very fabric of the city, where migrants live, what they can do and under what conditions encounters take place.

At the same time, the focus on migrant workers in this chapter also suggests that the urban settings of migration would appear to constitute the limits of governmental capacity to shape migration completely for its own ends. This is particularly apparent if we take seriously Simone's (2009) invitation to consider the uneven spatio-temporal terrain of urban life and the constitution of peripheral spaces as part of the differential inclusion of migrant subjects. The production of peripheral urban spaces is clearly critical to the management of migration, particularly temporary migrants, which rests on non-citizens being incorporated into labour markets but not invited into the normative possibilities of citizenship. The periphery allows for migrants to be placed at arm's length from full inclusion in society. This may occur in a literal geographical sense as in the case of Seoul but it is crucial to recognise that the periphery is a more complex assemblage than this and may not be based on absolute distance from the urban core.

There are three key issues that emerge from this chapter. First, it is clear that a key component of the urban politics of migration is the careful negotiation of visibility and invisibility in daily life. For the most part migrants desire and seek out invisibility, they seek not to foreground their migrant status, not to highlight their transgressions of governmental regimes and not to draw attention to their ordinary practices and lives. The peripheralisation of migrant worker lives serves this purpose relatively well, providing shelter from the close attention of the state, institutions, employers or indeed family back home. Yet, the invisibility generated in the periphery is also clearly problematic in that it shrouds the vulnerable position of some migrants, the exploitative practices of employers and the contradictions between legally established rights and the actualisation of conditions of existence. In this regard, visibility is hence also an important if transient tactic in migrant lives. In order to generate new settings in which otherwise silenced subjects can be seen and can speak, migrants must become visible, they must generate recognition. This can occur in multiple ways including working with institutions, forming co-national and multinational organisations or taking political action in the urban core in ways that highlight both the circumstances of migration and the capacity of migrants to claim rights.

Second, this discussion of the politics of migration in migrant worker lives also suggests there is a need to pay careful attention to the spatio-temporal configuration of the urban lives of migrants, not to presume that they live *in* static locations but rather to recognise migrant lives as active constituents of the urban spaces they inhabit. Workplaces, neighbourhoods, markets, places of worship, streetscapes, accommodation, parks and other places are not simply experienced by migrants as supposedly temporary residents but are actively, if not always intentionally, remade in the process. Even in the urban periphery, however, migrant agency is not equally distributed in the production of space. Rather, in some spaces and times, such as the regimented factory floor, migrants must submit to the times of others both in terms of duration of working day and intensity of labour during that time. Other spaces, perhaps even the nearby spaces of factory-based accommodation provide opportunities for slower time, for the generation of collective feelings and a mobile commons that supports a more meaningful life abroad.

Finally, then, drawing attention to the urban politics of migrant lives highlights the ongoing tension between a governmental desire for control over mobile bodies and the desire of migrants to control their own mobility. While the intricate biopolitical techniques of migration management can control the border in ever intensive and extensive ways, particularly in an effectively island state like South Korea, these same strategies run up against the messiness of everyday life, of workplaces, of public spaces and of the desire for migrants to direct their own destinies. The figure of the unauthorised migrant maintaining a freedom to move seems the most significant here but in not submitting completely to differential inclusion, integration or assimilation imperatives in their daily lives all migrants articulate a will to control their own destiny to shape their own mobility in ways that exceed what is expected of them. This is a theme that will continue in the chapters that follow as I trace the discrepant migrations and urban lives of international students and English teachers. In Chapter 5 the focus turns to international students and the ways in which desiring migration is channelled towards productive forms of diversity on the campus and in Chapter 6 to the alignment of whiteness as privilege and the incongruent migrations of English teachers.

Acknowledgements

Portions of this chapter have been drawn from Collins, F.L. (2016). Migration, the urban periphery, and the politics of migrant lives. *Antipode*, 48(5), 1167–1186. doi: 10.1111/anti.12255.

Endnotes

1 Gwang-lip, M. (2008). Expat's death prompts activism. *Korea Joongang Daily*. Available at: http://koreajoongangdaily.joins.com/news/article/article.aspx?aid=2887557 [accessed 14 Dec 2017].
2 Shin-who, K. and Si-soo, P. (2010). New drug test plan angers native English teachers. *The Korea Times*. Available at: http://www.koreatimes.co.kr/www/news/nation/2010/07/117_69263.html [accessed 18 Dec 2017].
3 The MTU was finally legalised in June 2015. Kyung-mi, L. (2015). After 10 year struggle, Migrants' Trade Union wins official status. *The Hankyoreh*. Available at: http://english.hani.co.kr/arti/english_edition/e_international/697696.html [accessed 14 Dec 2017].

Chapter Five
Channelling Desire and Diversity

I want to capitalise on that. I want to capitalise that um, because the – the core idea of this scholarship was that they want to make regional leaders, for [this Korean company] in India. Or other parts of the world, like Russia, or – developing economies. They want to have regional leaders, so as to present these leaders in those countries, and probably capitalise in terms of business. So, I do feel it's very important for me that I need to – not only live up to that, but – I mean, if they're capitalising on me, I need to capitalise on this. You know, it's like mutual capitalisation actually. (Nardev, India, Male, Student SNU)

Nardev's narrative of coming to Seoul to study engineering at Seoul National University (SNU) epitomises commonly held presumptions about ways in which international student migration is invested with aspiration and privilege. His parents, a university professor and school teacher, had long invested in their two sons' imaginative possibilities for the future, emphasising education as 'something that is most important in one's life'. Nardev had graduated from a 'premier' Indian Institute of Technology and went on to work for a major Korean technology company after graduation where he was successful in making a name for himself in research and development through patents and publications. Like his older brother, Nardev also wanted to study abroad and was successful in receiving a very prestigious global scholarship offered by a major Korean conglomerate that fully funded him to complete a master's degree in computer engineering at SNU and provided guaranteed employment for three years following graduation. As his excerpt above suggests, Nardev conceived of himself as a future regional or

Global Asian City: Migration, Desire and the Politics of Encounter in 21st Century Seoul, First Edition. Francis L. Collins.
© 2018 John Wiley & Sons Ltd. Published 2018 by John Wiley & Sons Ltd.

global leader, embodying exactly those desirable traits that are regularly espoused by policy makers promoting high skilled or talent migration (Shin & Choi 2015).

Nardev's narrative, and that of some other student participants, echoes the discursive framing of international students as 'kinetic' aspirational subjects who are presumed to have substantial capabilities and desires to move across borders (Sidhu et al. 2016). Johanna Waters (2012: 128), for example, one of the leading contributors to this field of work, goes so far as to suggest that 'it is axiomatic to state that internationally mobile students are invariably privileged'. From this perspective, privilege has provided a key explanation about the cultivation of desire for overseas qualifications in the reproduction of class and the interest in social and cultural distinction as well as the capability of students and their families to financially support global mobility (Findlay et al. 2012). These notions of the international student as a highly mobile and aspirational subject are also apparent in both institutional and state efforts to attract and retain mobile students (Kono & Chang 2014). In South Korea and other parts of East Asia, for example, the recruitment of international students has been regularly tied to both the attainment of 'world class' status for universities (Moon 2016; Yang 2016) as well as a view of students as potential human capital for addressing skills gaps and concerns around demographic renewal (Shin & Choi 2015).

The claim that international student migration is a process particularly invested with desire, more so than labour migration for example, is problematic in light of the discussions in the preceding chapters. Indeed, we have already seen in Chapter 3 that purportedly low-skilled labour migration is also generated in relation to its imaginative possibilities and in Chapter 4 the ways in which remaining in Seoul even as an undocumented migrant and progressing a politics of recognition in the periphery are fuelled by desires that have both instigated and sustained migration's transformative promise. As Cheng (2011) insists in her account of women on American military bases we need to recognise all kinds of migrants as not only objects of desire but also subjects whose own desiring reveals agentive will in relation to migration, daily life and the maintenance and transformation of self.

Nonetheless, there is a clear contrast between the ways in which international students such as Nardev are framed discursively as present and future subjects and both labour migrants and English teachers whose role it is to fill current occupational niches. Unlike workers and teachers, international students are not only viewed as desiring subjects but also regularly understood as desirable, not only for their immediate contribution to university and urban life but also their future role in economy and society. Accordingly, the priority for policymakers and institutions has been to capture and channel flows of international students and 'capitalise' on their presence for future benefit. The contrast to migrant workers who are the target of management and exclusion strategies, and English teachers who are regulated in more ambivalent but nonetheless controlled ways, reveals how the varying desirability of populations takes shape in the active production of stratified migrant categories. As Hoang (2017: 2) notes, 'disciplinary tactics and technologies [are] deployed to entice the "best and brightest" foreigners to come and stay as well as to keep settlement and membership prospects beyond the reach of less "desirable" populations'.

This chapter takes up this question of the production of migrant categories in relation to the ways in which forms of desiring-migration are subject to territorialisation or stabilisation as individual migrants become part of urban life. I focus primarily but not exclusively here on international students because while they are often viewed as 'kinetic' subjects who can move easily across borders and adapt to a wide variety of settings, international student migrations are also enabled and contained in their incorporation into campus and urban life through strategies deployed by the state, institutions and social networks. These strategies are key to creating the migrant category of international student and defining its appropriate place in the city and campus, vis-à-vis both domestic students and other migrant groups. The accounts here point to students encountering segmented spaces on campuses where their presence as international students is channelled in ways that seek to ease their mobility but that in doing so also distinguish them from domestic students and other local residents. Beyond campus boundaries, students can have relative freedom but as the discussion here demonstrates, encounters in the city are unevenly experienced and nationality and gender in particular are brought to the fore in shaping what relations can be established in urban life. These insights, then, advance the argument established in Chapter 4 about the differential geographies of migrant life to reveal the ways in which the particular positions of different migrants in Seoul are assembled through a wide variety of strategies for structuring and containing migrants within discrepant schemas.

5.1   Territorialising Migration in the City

Chapter 4 demonstrated some of the key discrepancies that exist in the urban lives and geographies of different migrants in Seoul. Most notably the chapter revealed the peripherality of migrant worker lives and the striking contrasts that exist in relation to the geographies of English teachers in the suburban spaces of Seoul and international students living adjacent to the spaces of the city's major universities. These geographies reveal some of the ways in which the patterns of migration take shape in relation to the existing constitution of Seoul – the development and outward movement of manufacturing districts, the distribution of English language academies in middle class residential zones and the concentration of prestige and reputation in an era of globalising higher education. In the language of assemblage urbanism, Seoul is not the sum of these and other parts but is constituted through particular types of co-functioning, something that is clearly revealed as different migrant lives come to make sense in these different parts of the city.

The value of this focus on assemblage in discussing migration and cities is that it makes it possible to prioritise the relations *between entities* rather than to the arrangement as a whole. When it comes to cities, this means recognising that while there is a certain level of autonomy between the entities that make up the city – its buildings, people, governmental orders, images etc. – simply adding up these entities or new additions or arrivals can never explain the whole. Rather, it is necessary to focus on what DeLanda (2006) captures in the term 'relations of exteriority', the effects that are generated because of *the particular way they are aligned*. As the narratives in this chapter reveal, the presence and practices of international students, as well as teachers and workers, does not just occur but rather is subject to different disciplinary tactics – in the infrastructures that support or manage arrival, in housing, study and work arrangements and in the variable timing of migrant lives (the latter of which is discussed in more detail in Chapter 6). My argument is that these infrastructures and arrangements seek to capture and channel the desires that undergird the migration of students, workers and teachers and in doing so contribute to the varying desirability of and opportunities available to these migrants in Seoul. It is here that we see the articulation of movement and urban life:

> So as physical and political infrastructures stratify movement into different interactional possibilities, and steer people into specific densities and speeds, as well as open weakly controlled reverberations, urban life makes itself known in various ways. (Simone 2011: 356)

Put otherwise, the coherence of particular migrant categories is not only a result of the migration regime and transnational migration infrastructures, but also takes shape in processes of emplacement including the specific ways in which migrants become part of different urban spaces (Lin et al. 2017). Often the role of the state is less apparent in these instances as employers, industry organisations, accommodation providers or educational institutions amongst others come to play a more significant role in arranging the day-to-day lives of migrants. While often viewed as simply the mundane everyday features of managing the presence of migrant populations these techniques are critical to establishing the conditions of migrant life and in separating out or reinforcing differences between migrant groups (Hoang 2017). This is already apparent in relation to the peripheral lives of migrant workers discussed in Chapter 4, a point made also in other research on labour migration where the alignment of housing provision, employment and social time and activities can serve as a form of 'social quarantining' (Horgan & Liinamaa 2017). In contrast, the arrangement of privileged housing and transportation packages act as an infrastructure for supporting the very different characteristics of everyday life for highly skilled 'expatriates' (Cranston 2017). These mediations, then, are part of the wider intensification and diversification of strategies of control in relation to migration (Geiger & Pécoud 2013) that link the transnational movement of peoples to their position in everyday life and urban spaces.

When it comes to international students, universities play a considerable role in shaping the mobilities of these migrants and the ways in which they are incorporated into campus spaces and urban contexts. Both KU and SNU have pursued aggressive globalisation strategies over recent decades where the recruitment of international students has been seen as a core component of achieving world-class status. This includes efforts to rebrand universities, the introduction of English Medium Instruction (EMI), hiring overseas faculty, developing international summer schools and student exchange programmes, recruiting degree-seeking international students and building international networks (Collins & Park 2016; Sidhu et al. 2016). These initiatives often hinge on ideals of diverse campus life and global encounters wherein international students are both symbolic and embodied resources for manifesting globalisation. Such ideals are difficult to engineer, however, and universities engage in globalisation initiatives with a range of ideologies about ethnic difference and interaction. Moon (2016), for example, notes that while there have been proactive efforts to recruit foreign students into leading institutions in South Korea, there has been much less willingness to allow their presence to alter campus life. She suggests instead, 'notions of ethnic nationalism remain firmly entrenched at the level of university curricula and also at the level of everyday interactions between foreign and local students' (Moon 2016: 92). As the discussion in this chapter will demonstrate there are also a range of initiatives on campus that seek to contain international students, providing specific spaces and services, often in English rather than Korean, that support the presence of these students but in doing so also separate them from domestic students. More recently, the Korean Ministry of Education has indicated that it would push to make legislative changes to allow universities to create specialised departments and programmes that cater *only* for international students (Bothwell 2015). While envisioned as part of the making of global campuses, and materially having this effect through the circulation of international students, such initiatives also create highly demarcated spaces that capture and channel student migration.

A wider set of concerns about supporting and enhancing the migration of international students emerge beyond campus spaces in the uneven geographies of the city. Scholars have often noted, for example, that international students have fewer encounters with domestic students and local residents than they expect and tend to socialise either amongst co-nationals or with other international students (Jones 2013; Brooks & Waters 2011). There are often multiple reasons for these disruptions to the expected cosmopolitan-ness of student migration, including the different language capabilities of students, different desires for encounter expressed by international and domestic students, cultural differences around socialising and discrimination (Collins et al. 2014). In broader urban spaces too there can be forms of segregation that emerge through constrained social networks or a range of intermediary and institutional practices. Fincher and Shaw (2009), for example, observe forms of 'unintentional segregation' in housing that has been generated by institutional efforts to cater for international students in Melbourne. Similarly, in my earlier research in Auckland unplanned alignments between institutions, local government, property developers, education agents and rental markets channelled students towards inner city apartments where encounters with others were minimised (Collins 2010).

The end result of these patterns of campus and urban life is that overseas education can manifest in segmented ways where the claimed benefits of migration for both international and domestic students, as well as the reconfiguration of university campuses and local contexts, can be much less than expected (Sidhu et al. 2016). The effort to capture and channel the desiring-migration of international students and other migrants towards particular outcomes is a fraught undertaking. Chapter 4 has already revealed this in relation to migrant workers and the ways in which the desire to direct one's own mobilities and life in the city can substantially exceed expectations of control that are created in migration regimes. While the articulation of migration and urban life reveals the exercise of power, rationality and intelligibility by various actors seeking to organise movement there is always the possibility for alternative unexpected arrangements. The focus on desire and assemblage reveal this because they do not treat individual migrants as only examples of wider migration categories but rather address the multiple dimensions to mobility and subjectivity and the alignments that emerge with places. In this context, even governments and institutions promoting migration 'cannot anticipate the problems that will ensue as a result' (Ong 2005: 349) of particular alignments but must rather respond to unpredictable developments and reconfigurations. As we shall see, international students may be valorised as desirable subjects but their own desiring-migration means that even intensive efforts at control reach their limits.

5.2   Infrastructures of Arrival

One of the key infrastructures involved in establishing the place of migrants in the city and in relation to other urban residents are the various programmes in place for coordinating arrival and emplacement. Infrastructures of arrival can involve highly managed processes of escorting migrants into and through airports and placing them in particular locations as well as looser sets of guidelines and information provided within compulsory or voluntary orientation programmes. As part of the wider 'physical and organisational architectures' (Lin et al. 2017: 168) of migration arrival infrastructures of this kind matter because they set expectations for migrants, coordinate movements and relationships and give meaning to migration in ways that 'variably produce migrant categories that are often now taken for granted'. At the same time, arrival infrastructures sometimes also vary within migrant categories and can as a result produce variable experiences of mobility that are shrouded by labels like migrant worker, international student and English teacher.

Two broad patterns of arrival emerged in the narratives of international students in this research. On the one hand, students on government, corporate and full university scholarships (18 out of 40 participants) described a pattern of arrival and emplacement within Seoul that was managed by administrative processes that were beyond their control. As an example, it is worth returning to Nardev (India, Male, Student SNU) because he was on an extremely prestigious global scholarship that was targeted at young people in China, India and Russia who had talent in electrical engineering and computer science. The scholarship determined the university he would study at, the courses he would take, the extra-curricular activities he should be involved in and the housing he would be provided. For Nardev, this meant that:

> Everything was arranged. We just had to move into – we had to take flights. I mean, they gave us the flight tickets, and we just had to come in here, and the accommodation was already there. Yeah, everything was pre – everything was done. I mean, someone was there to receive [us] at the airport, so we had no hassle to move in. It was a very easy move in.

Unsurprisingly, Nardev and other participants in similar situations viewed these arrangements very positively; they served to smooth the friction of mobility and overcome challenges associated with moving into an unfamiliar learning and urban environment. While Nardev was one of only two students receiving corporate scholarships, other students arriving through government funding reported similar levels of support and channelling of their migration and education pathways. Students in the Korean Government Scholarship Program (KGS), for example, entered South Korea first to study Korean language for 12 months at an institution selected by the National Institute for International Education and Development (NIIED) before applying for degree programmes. Students were provided with full tuition fees and living costs for language and degree study, round trip travel and living costs as well as being provided with airport pickup, orientation and pre-arranged accommodation. For the five KGS students in this research, these arrangements eased arrival processes and reduced the uncertainty associated with studying in a country where they did not understand the language or have familiarity with the culture.

These linkages between the scholarships and the channelling of mobility and emplacement were also apparent in the case of the five Malaysian students, who had all come to study in Seoul through government-sponsored scholarships. Their migration was situated within the Malaysian government's 'Look East Policy', an initiative established by former Malaysian prime minister Mahathir Mohamad in the early 1980s to learn from Japanese and Korean economic development (Milne 1986). In addition to foreign direct investment and technology transfer Look East has provided scholarships for students to study in Korea since 1984 with a particular focus on engineering and other technical qualifications. While the numbers in the programme remain small (around 80 per year), the students who enter are provided full scholarships for fees and living expenses and are provided a range of social and cultural support through the Malaysian Embassy in Seoul and *Persatuan Pelajar Malaysia Korea* (PPMK), the Malaysian students association in the Republic of Korea. Students arrive in Seoul in cohort groups, undertake a two year diploma at a technical college in Seoul and then select from participating engineering programmes at three major universities. Their education and life pathways are guided not only by the formal scholarship requirements but also by a well-established network of students in the PPMK who advise newer students towards particular choices. For Kelvin (Malaysia, Male, Student KU), for example, this advice led him to KU because it had a larger number of English language lectures, and because at [another university] 'the seniors have or weren't doing that well' in terms of grades. He described the quite detailed arrangements for support:

[Enrolment] was done by the student society in the uni because the student society have this education bureau and this education bureau is supposed to be in charge of all this enrolment stuff so what they did was they collected all the forms passed it to the students and the students will have to fill it in and the students will pass up the form themselves to the administration office of the University....You can actually seek help directly from the student society or you can just give one of your seniors a call or message just message them and they will help out. They are quite friendly.

These systems of arrival and support are indicative of the kinds of 'migration infrastructure' described by Xiang and Lindquist (2014). They have the effect of modulating migration such that international students in these programmes are not simply coming to study in South Korea or Seoul but in specific universities and programmes and in their wider lives also in particular housing arrangements and social opportunities. They also create privileged spaces of circulation where *some* international students who are deemed by governments or corporations to be talented or to hold potential experience enhanced mobility, both transnationally and in their adaptation to life. They become 'kinetic' subjects in migration and orientation to campus life but only as a result of strategies for facilitating their presence; their status and capacities in the city are not only their own but are rather produced through the relations that are established for them.

Students arriving outside of these structured programmes provided quite different accounts of their migration and emplacement. Rather than relying on corporate and government-sponsored guidance, they were faced with the task of navigating their way through complex visa processing, enrolment, arrival and housing. Some students turned to intermediaries for assistance, such as Mingyu

(China, Female, Student SNU), who came first to KU's language programme before transferring to humanities at SNU:

> When I first came to Korea, it was with the help of an agent. It was a group organised by the agent who came together, so I already got to know one or two people before I left China. So I wasn't one of those who was alone here who didn't know anyone when I first arrived. On the one hand, I had the help of seniors, and on the other hand, I had friends with me, so the difficulties that I met weren't that many.

As Mingyu suggests, pre-arranged travel, transport from the airport and accommodation was important not only for avoiding 'difficulties' but also for establishing connections with other people arriving as students. In other instances, students described using connections with people in South Korea, seniors, alumni of their university or school or even family to assist with arrival.

> When I first came here, because I had a senior who was studying here. When I was in the midst of the application procedures, she helped me out as well. When I arrived, I took the airport bus from the airport to Seoul National University. But after I got off the bus, my senior came to fetch me. I stayed in her dormitory on the night of my arrival. The next day, she brought me out to get a mobile phone and to look for accommodation. (Yunru, China, Female, Student SNU)

Like those participants who were funded by scholarships these connections were important for easing arrival processes and establishing basic elements of life in Seoul: accommodation, social contact, an orientation to the city. They did differ, however, in terms of their effect on where and with whom students lived in ways that influence the encounters participants had in the city. Whereas scholarship students were placed in university dormitories amongst a diverse range of students, those who arrived independently or who were supported by education agencies were often quickly settled into co-national accommodation. In our research, these students were also overwhelmingly from China[1] and as a result were able to utilise well-established social networks to find off-campus housing:

> Actually, I lived with another Chinese girl and I didn't move out until this year. She needed another roommate to share the place. Some friends introduced her to me when I was in China. I just met these people online. I asked them how to find a place to live and they were very helpful. Some of them told me this girl needs a roommate. Actually, many students who come here to study use that website and there are many of this kind of websites. (Huifen, China, Female, Student KU)

As Huifen suggests, websites as well as other social connections serve as an important support for students who do not receive directed advice from institutions or who are not funded by scholarships. These infrastructures enable as well as guide mobility because they allow students to reduce some of the unpredictability involved in travelling across borders and undertaking study in what was often a very unfamiliar environment.

It is useful to contrast these differing international student mobilities with the arrival experiences and infrastructures of migrant workers and English teachers. Arrival infrastructures were particularly prominent in the narratives of people migrating through the EPS regime but took on a very different character from those narrated by either scholarship or non-scholarship students. Participants spoke of travelling in groups from their home countries, in some cases being asked to wear uniforms and name tags, and being met upon arrival at Incheon International Airport and taken as a group through immigration formalities and then by bus to training centres for 3–4 day introductions to language, cultural norms, labour laws and workplace skills. As Monty (Philippines, Transgender, EPS) put it, such programmes involved a clear intent to generate compliance: 'They introduce you to Korean culture. Then they kind of brainwash you, to be good to your bosses'. Following the completion of orientation programmes, workers would be collected by employers and taken to their usually co-located accommodation and workplaces.

English teachers who had been recruited within government run programmes such as the Gyeonggi English Program in Korea (GEPIK) were not subject to such strict controls but were picked up at the airport and required to take part in an 8–10 day orientation programme that took place the day after arrival. This programme provides teacher training and introductions to cultural differences, social network formation and study or teaching skills. Like international students the orientations offered to teachers within GEPIK and other programmes was generally viewed positively:

> This orientation we had a really great…had a couple of friends, they gave great lectures and presentations about Korean culture, about transition, there is support for people who, you know what I mean. (Bethany, USA, Female, English teacher)

For the majority of English teachers who worked for private English language academies there was usually no formal orientation process. Instead, teachers were often but not always picked up by the employer or an arranged taxi and taken directly to their pre-arranged accommodation. In many instances, they were expected to start teaching either the next day or after a few days of observation. While co-national networks then were not immediately part of arrival processes, many sought assistance from foreign English teacher colleagues, sometimes Korean English teacher colleagues or in the internet communities established by English teachers in Seoul and other parts of South Korea. Like students arriving outside of scholarship programmes their migration appeared more turbulent, characterised by both freedom but also uncertainty that will be discussed in greater detail in Chapter 6.

While the content, and levels of freedom and unfreedom, of these arrival infrastructures clearly varies considerably we can read each as examples of the ways in which migrant categories are produced, both in the form and direction of their migration but also in relation to their place in the city. In each case, these strategies also seek to produce particular kinds of migrant subjects, compliant workers, well-adjusted teachers in public schools and high-performing international students who will not be hindered by the friction of migration. In doing so they also establish expectations for the kinds of encounters that different migrants will have as they arrive and in the work, study and urban spaces they will inhabit: EPS workers are oriented around workplaces and employment relations, some English teachers learn about cultural differences and gain opportunities to meet other teachers in their area and some international students are guided smoothly into the opportunities offered by Korean higher education system. As part of mechanisms for managing migration, such programmes acclimatise migrants to be different in their lives, working with and extending the discrepancies established in the regimes that govern different forms of migration. They produce migrant categories *and* their place in the city, a pattern that in the case of international students also extends into the organisation and experience of campus spaces.

5.3   'Everything is Within the School Campus'

The patterns of mobility and emplacement described by international students meant that they experienced considerable constraint in their initial encounters with urban life in Seoul. Having arrived through relatively structured educational pathways established by governments, institutions, intermediaries or co-national social networks there was a strong pattern amongst students of describing being 'trapped', 'closed in', in 'another world', or staying in 'my little corner' of the campus. This was particularly apparent in students' accounts of their first few months in Seoul. Vitor (Brazil, Male, Student SNU) provided a good example of this. His migration was supported by a KGS scholarship and he had been assigned initially to study at the language institute at SNU and provided with accommodation in a dormitory on campus. Having no knowledge of Korean language or South Korea before arrival, he described a sense of being guided into an environment where there was little possibility of meeting people outside the KGS scheme or even to become accustomed to life in Seoul:

> Because, I think that my, my first months here, they were so hard, because I was trapped at the environment. So, I couldn't meet people from – different people, it was very hard for me. We were closed in a group, and everybody was suffering! Because of the language, because of everything. So, yeah. It was hard for everybody, and maybe if I, I dunno. If I went to meet different people, if I had not – I, I dunno. I would change my attitude, the first, first few months that I was, I was here.

The feeling of being 'trapped' and 'closed in a group' reflects the manner in which scholarship students like Vitor were supported into facilities that were tailored for international students. Indeed, rather than a sense of exclusion from the university, the narratives of students like Vitor revealed a carefully guided inclusion of international students into specifically *international* spaces within the existing university structure and campus environment. This is particularly the case for scholarship students who were supported by formal programmes but similar effects emerged through co-national networks; both forms of infrastructure play the role of facilitating migration but also stabilising its outcomes in particular material and imaginative spaces not only within the city but literally within particular parts of the campus where international student lives are viable.

At both KU and SNU the growing focus on recruiting international students has been accompanied by the establishment of a whole range of new campus forms: new English medium curriculum, language education programmes, international summer schools, international student centres, student and staff housing, English-speaking administrative staff and new institutional paraphernalia like websites, brochures and events that declare the global intent of these universities. At KU the construction of 'Dongwon Global Leadership Hall' and 'CJ International House', for example, reflect attempts to cater to the needs of international and exchange students in terms of administrative processes of the university, learning support and housing expectations. SNU has developed similar facilities, such as 'CJ International Center' that includes a 'one-stop service center' to support international students and has expanded its provision of on-campus housing with priority given to international students. In both cases, the facilities are staffed with English speaking administrators who can provide targeted advice and support on housing and educational issues as well as serving as spaces for supporting the development of social networks between international students. These developments, then, both signal the shift to globalising higher education through international student mobilities as well as serving as key sites for managing those flows and their outcomes (Sidhu et al. 2016).

These attempts to reconfigure nationally oriented universities to globally mobile students serve in important ways to support internationalisation. They make it possible for students with no prior familiarity with South Korea (such as Vitor and many others in this research) to study a degree programme at institutions like KU and SNU. At the same time, however, they also tend to operate in a way that minimises inconvenience for international students and in the process separate them from Korean students in spaces of housing, administration, study and everyday life.

> I live in the students' dormitory organised by the university. But it is very expensive and the living quality is not that great. But to be honest, with that price in Seoul, I am satisfied with my current living situation. I live with another two Chinese

students. ... I applied for the place but the university just randomly picked up room-mates for us. It is actually very hard for people to get into students' dormitories. Since many people apply for it every year. ...They are all international students here, from China and other countries. Everyone is friendly. (Qianfan, Male, Student KU)

Qianfan's account of dormitory housing was indicative of wider patterns of on-campus housing experiences amongst international students. While some were housed alongside Korean students, or even with Korean roommates, many also described a pattern of interaction in dormitories that was bifurcated along international–domestic lines and then at times demarcated amongst international students in terms of nationality.

These housing arrangements played an important role in establishing patterns of socialising where language and nationality came to dominate many students' experiences. Participants spoke about 'staying together as interna-tional students' (Priya, Sri Lanka, Female, Student SNU), a pattern that reflected shared circumstances but also the reality that many international students came to Seoul with little or no Korean language ability. In this con-text, on-campus housing, or arranged housing off-campus with co-nationals was considered the 'easier' option:

I think staying in dorm is easier for me to make foreign student friends, not Korean friends! ... I think it's easier for international students to mix together. Because like, international student, they more like to make friends. (Jing Yi, Malaysia, Female, Student SNU)

It is important to emphasise that many participants in this research both appreciated the sense that housing and internationally-oriented administrative services made their lives 'easier' while also noting the effects it had in terms of limiting opportunities to build relationships with Korean students. There was a sense that social networks in particular were constrained to relatively small social circles and that for many students their lives were largely contained within particular spaces within the campus.

I think it [socialising circle] is very small. For example, my life has been centred around my university campus. I don't even leave the campus for one or two months. Since where I do sports is located within the university campus and whom I socialise are mostly working within my lab or my Chinese friends studying here. (Jiahui, China, Female, Student KU)

The characterisation of life as being centred around the university campus expressed by Jiahui and many other students also extended in important ways into courses of study and classroom experiences. As noted above, only very few students in this research had proficiency in Korean language before arrival. As a result, students were confronted with difficult choices in terms of course enrolment: becoming proficient in Korean language, trying to take only English language courses or mixing Korean and English courses once proficiency was established. The first pattern was followed by students from China (including Korean-Chinese) in this research, many of whom reported being more comfortable learning Korean than undertaking classes in English. While this meant these students eventually studied alongside Korean students, their initial campus and social patterns were established in a largely international or even co-national classroom environment:

> At that time [of first arrival], there were a lot of foreign students from China in my discipline. So, it feels like we came to another world and we go to class together. It was quite interesting.... Nothing too inconvenient... No problems like culture shocks and things like that.... But for me, I was surrounded by many foreign students, especially students from China, so I didn't feel lonely. (Huizhong, China, Female, Student SNU)

Studying alongside other international students, especially those from the same national or language background, clearly smoothens the experiences of international students adjusting to university life. As Huizhong describes it, there is little sense of 'culture shock' and relatively little feeling of being 'lonely'. At the same time, however, read across international student narratives and the effects of these patterns manifest in campus space it is apparent that globalisation of higher education can take on a highly contained form. As universities seek to recruit and retain international students and to provide them with a 'convenient' experience, and students themselves seek out familiar social opportunities, levels of segmentation emerge that seem to question both the cosmopolitan promise of higher education and the kinetic capacities of international students.

For other students who are already proficient in English, part of the reason they have chosen to study at these universities is the availability of English language courses. This was particularly the case at KU where over 40% of courses are offered in English, but students at SNU also discussed targeting their course selection to English-only courses when they could to ease anxieties about studying in Korean.

> The semesters I've had, all for my major, I've taken English courses only. But eventually I will have to take Korean classes. You can't choose. There's only so much that open up in one semester that we have to keep to those. So. Like – I think like it doesn't matter like how good my Korean gets, it's just that university level, that makes it that much difficult. (Arturo, Mexico, Male, Student SNU)

Enrolling in English-medium classes addresses the concerns that some students have about the difficulty of studying in South Korea. Indeed, it is arguably the very presence of these courses that makes it possible for some students to imagine studying in South Korea in the first place because it constructs an explicitly international and English-centred environment.

> During first year I took all the English classes, so my class is not a big problem. But mixing with the Koreans is a big problem, because you have to speak Koreans, and they don't speak English at all. (Jingyi, Malaysia, Female, Student SNU)

As Jingyi suggests, enrolling in English classes makes it easier for many international students to progress in their studies but also means that they have limited opportunities to meet Korean students and build relationships. While EMI courses will have some Korean students, especially at KU where the proportion of EMI is higher, the level of interaction amongst students was reported as being much lower. In effect, many students in this research reported studying in a parallel world to Korean students who would interact with professors in Korean even in EMI lectures and who would not feel comfortable conversing casually in English. As Priya (Sri Lanka, Female, Student SNU) suggests, this can stifle even well-established desires for interaction:

> I really want to do some extra-curricular activity like a sport or a club, but I feel like my Korean language is going to be a problem, so I haven't, I haven't, I'm not in any club or anything. But I, I want to do it next, next year. But, I try to go into like the activities organised by SISA, which is the SNU International Students Association. I try to go for their activities.

Desiring the acquisition of capacities to speak a language and the opportunity to meet others are not easily enabled, regardless of the sorts of interventions that universities or others undertake to create globally oriented campuses. Indeed, they can have the opposite effect in territorialising and containing student lives through a combination of arrival patterns, language differences and course enrolments that mean international students can experience considerable restrictions within the space of the campus. Unlike EPS workers and English teachers their day-to-day lives are positioned alongside Korean students who are nominally their peers. However, as the excerpts here suggest, arrival and emplacement in campus spaces is often experienced as a highly structured process and one that seems to minimise interaction with Korean students and direct international students towards the globalising rather than the still national spaces of universities like KU and SNU. International students are undoubtedly desired for the way their circulation embodies globalisation and they themselves are pursuing what they see as *international* education opportunities in Seoul. Yet, at the same time, their movement is also subject to institutional and personal logics of containment that work through an architecture of social difference where national and global can be maintained side-by-side. As Section 5.4 demonstrates, these uneven experiences of campus spaces also articulate into quite different sorts of urban lives for international students.

5.4   Encountering Seoul

> I feel like I'm a forest monster. It's as if I have been separated from the rest of the world. Therefore, sometimes I visit Myeongdong[2] because I'm afraid of staying here for too long a time and as a result, become stupid. So I must go out often. (Minsheng, China, Male, Student SNU)

As Minsheng suggests, one effect of the campus settings described above is to generate a feeling of isolation that can undermine the purported objectives of international student mobility, to learn and explore different places and opportunities (Brooks & Waters 2011). Indeed, while qualifications acquired through study in Seoul are clearly valuable, *international* study also hinges on forms of social and cultural capital beyond the institution (Waters 2008). While obviously in jest, the idea of becoming 'stupid' expressed by Minsheng speaks to the manner that learning takes place in situated ways that involve classroom, campus and wider encounters (Lave & Wenger 1991). This insight also speaks to the manner in which migration more generally takes place across diverse geographical terrain, beyond the specific sites, workplaces and universities, where migrant lives are centred.

One of the insights offered in the focus on urban assemblages, however, is to recognise that cities are not composed of distinct mutually exclusive spaces – residential areas, manufacturing districts, university campuses – but rather it is relations between bodies, objects and ideas that matter as much in assembling urban life. Indeed, as Chapter 4 has already demonstrated socio-legal status and the geographies of migrant life shape encounters with the city in ways that exceed the parcelling out of different spaces. The ability that different migrants have to navigate and negotiate city spaces, to encounter Korean populations and to build meaningful attachment and relations with people and place varies considerably between workers, teachers and students as well as within these groups. These differences reflect the territorialisation of urban assemblages, not only in formal categories and statuses, but also in the ways in which different migrant lives are organised spatially and temporally and the varying capacities that migrants carry with them – financial, social and embodied. Moreover, encounters in the city also mean that individual migrants are drawn out of the particular contexts where different migrant categories make sense. Students and other migrants can become more or less visible in their encounters in ways that can both empower individuals and serve as obstacles to desires for enhancing interaction with others and cultivating attachment to place. In this last section I draw attention to these wider urban lives and in particular the kinds of initiative that they demand and the uneven encounters that result for students.

5.4.1   *Taking initiative*

> It is comfortable to live here. Nobody knows me and I can do whatever I feel like. I can just mind my own business. When I need, I can meet a friend. You can't eat alone in China since everybody knows me there. But here, I can eat alone, I can wear whatever I like, I mean I feel very free and relaxed. (Qianfan, China, Male, Student KU)

As I have already suggested, many international students in this research found their lives channelled towards relatively confined campus spaces that aligned with the lives of other international students, often those from the same country. For Qianfan, this configuration of life abroad was not something that was identified as particularly concerning. Indeed, he noted that his mobility to study in Seoul was not the result of a specific interest in South Korea or studying overseas but rather because he could not get into the university he wanted to in China. He spoke some Korean because he grew up in a Korean-Chinese family but saw himself as Chinese and emphasised that his main purpose was to study and gain a qualification. In other words, his migration was situated in relation to desiring educational progression and the possibilities that are often promised in university credentials rather than the oft claimed cosmopolitan potential of student migration.

This was the case for some students in this research – study was viewed as a time-limited activity for credential acquisition and life beyond the campus was not viewed as critically important. Similar accounts also emerged in the narratives of some workers who placed emphasis on saving money and living a quiet life outside their work schedules. Chati (Thai, Male, EPS worker), for example, relished the quiet life of the urban periphery where his factory was located: 'It is peaceful and I can save money. I can enjoy the nature of Korea during weekends without using money. It is different from working in the city that everything is about money.' Qianfan also rationalised another component of his avoidance of wider social connections in the quote above when he highlights his comfort with the anonymity of life in Seoul: as Korean-Chinese he is rarely recognised as an outsider in day-to-day activities and as a result he has a sense of freedom and relaxation in the city. He also spoke about the limits of social relations that existed, however, and the way in which this revolved not only around his reception by others but also his own drive to interact:

> Some Chinese international students can get along with Korean students well. But I can't. Maybe it is my personality. *I don't take initiative to approach people.* Communicating with locals at a deep level has always been hard. It is not easy to make friends here. Plus, socialising has a cost, you need to spend money to do dinner and stuff.

Again, these expressions were not limited to international students but also emerged in different ways for workers and teachers. While some participants in all groups situated their lives in Seoul and South Korea as filled with the opportunity and excitement of building diverse relations and exploring new spaces, others sought more consolidated social interactions centred around the reproduction of familiarity. Karen (South Africa, Female, English teacher), for example, emphasised avoiding spaces where she might meet Koreans outside of work in a way that reflected a parcelling off of her working role as an English teacher from her social life: 'I teach English the whole day and I don't want to go home or go socialise with people who don't speak English'. Such impulses remind us that encounter is not a given in urban life but rather also revolves around desiring processes. For international students and other migrants, encountering others demands not only the possibility of being received in a positive fashion but also the interest in approaching others and the desires and capacities that might generate these interactions (Jones 2013). It demands a will to encounter and build relationships.

A useful comparison can be made between these varying accounts of avoidance and another Korean-Chinese student at KU, Liwei (China, Male, Student KU), who was studying in a very similar degree programme to Qianfan but expressed a very different orientation to life in Seoul. Unlike Qianfan's sense that overseas study was a route to only gaining a degree, Liwei described how he had 'wanted to go overseas for study ever since junior high school'. For him, overseas study was not simply about education or qualifications but rather about gaining more opportunities to encounter different ways of living: 'so I have been feeling that I want to experience new things and I want to live in a very international environment'. Desiring encounters with others are an important driver here of practices in the city itself. Qianfan has assembled a wide range of activities in his daily life, from meeting Korean friends to practice and improve his Korean, playing badminton with his Chinese friends and developing relationships with a wide range of international contacts:

> *I took the initiative* to make some friends when I studied in Sejong University. I went to the 'English corner' to practise my English at that time. I met lots of them there. People also thought I am an interesting person as an ethnic Korean-Chinese who also speaks English.... I just finished an internship four months ago and I got the information of that job from a friend here. My friends also tell me the culture and thinking patterns of Korean people.

In Liwei's experience, taking 'initiative' in meeting others through opportunities like 'English corner' or language exchange for Korean students clearly serves as a connection into wider possibilities. He has been able to build friendships across cultural difference, undertake an internship that was offered through one of those friendships and more generally develop a subjective orientation that supports further encounters. Indeed, in stark contrast to some other participants, including other Korean-Chinese, he argued that the 'whole atmosphere' in Seoul is positive, that 'Korean people in general are willing to talk to foreigners' and that there is considerable scope 'to be someone who is in-between'.

The desire that individuals have for encountering others, then, is something that can make it possible to establish connections into wider fields of social relationships and urban practices. It makes it possible for international students and other migrants to reconfigure the ways in which their lives are territorialised in the city by reworking relationships with people and place beyond the forces of containment that exist in the global university campus. As Bignall (2008: 131) puts it, while encounters often emerge initially by chance, desire can be a force for striving 'to become oriented in these relations – and indeed to actively forge and facilitate meetings – according to an active and affirmative will to power'. It is not surprising that some international students were oriented towards such striving, given the efforts involved in moving abroad for study and the goals of building relationships and learning about difference and making opportunities for their own futures. One example that was expressed by Liwei and several other students in this research was taking the initiative to find internships or part-time jobs that placed them in formal roles outside the university space.

> The company is recruiting us, in campus. Maybe they recruit all SKY [university students]. If I was in Gangneung [a small East Coast seaside town where Jeana initially studied], they never call! They say the global internship programme. … They only recruit the international students, 20 or 40 students, and a lot of nationalities. … Companies need the Chinese students more [so] almost 50% is Chinese. And one, two are – if there is hundred people, two Mongolian, or 50 Chinese, or 10 Japanese, like that. … From internship I can meet the master degree students. It was good for me. Usually the master degree students get the scholarship, or they have work experience, before, and – they totally enjoy the life in. They always call me the Friday night, let's hang out, go out, like that. (Jeana, Mongolia, Female, Student SNU)

Taking on an internship in this case, then, not only serves as a mechanism to become more embedded into different fields of life in Seoul beyond the campus but also clearly generates scope for more encounters. Jeana's internship was with a major Korean bank that is increasing its presence in other parts of Asia and accordingly is seeking to incorporate international students from key nationalities into the firm and to build its own connections for future market success. The alignment of students' own desire to learn and build opportunities, and their desirability in the eyes of a multinational bank forms a key component of assembling lives beyond their containment in the campus, not only for the present but also into after-study careers (Collins et al. 2017).

Moreover, like other students who had internships or jobs, Jeana also described how her friendship circles expanded through taking this initiative, it facilitated meetings that constituted a reconfiguration of her social connections into the assemblage of the city. This was particularly important for students who had fewer co-national or other networks when they arrived, including many of the scholarship students. For these students, building friendships demanded initiative and drive to connect with others. Danyiar (Uzbekistan, Male, Student SNU) explains:

> I was very social, like socialise, I like to communicate. So I tried to gather lots of friends together, and even if like A – if Friend A doesn't know Friend B, I'll want to make them friends too. So that's why like, I invite my friends to our gatherings, and make more and more group. And – you know, the networking is also important. Like in our life. Because in future, in 10 years, in 20 years, you don't know what will happen. Always you'll have like friends who can help you.

Friendship is here a kind of social assemblage that also forms relations with other parts of wider urban assemblages, it supports engagements with everyday life but it is also something that engenders pleasure (DeLanda 2006). While Danyiar talks about friendship in terms of its prospective value, the way in which meetings and connections might serve as an infrastructure for the future, he also emphasised his own interest and capacity to interact with Koreans and other international students:

> I like this country, you know. I like the people. So, I began like – me, as my – like I am very communicative. So, I found lots of Korean friends, so we began to communicate, and it was like, not difficult for us, as for me, adaptation.

Like all of the students discussed in this section, initiative and interest were key components in aligning themselves with life in Seoul beyond the campus. Mingyu (China, Female, Student SNU) provides one last example, where such initiative and interest is also accompanied by a willingness to undertake bodily alteration in order to align with social norms.

> Because I heard from the other seniors that, in the first few days of schools, I'd have to hang out and drink alcohol together to get to know the rest of the people. I'd have to go somewhere else and everyone will drink together in a room. I've heard of that quite early on. When I was in China, I never drank alcohol so after I came to Korea, I was told that, after I am admitted into the university, I would have to drink alcohol with others. So they taught me how to drink alcohol. It was [my friend] who taught me how to drink. ... After that, when I was hanging out with the Koreans, in the beginning, we were quite unfamiliar with one another while we were playing games and drinking alcohol. But as we drank more alcohol, we started to talk more. To become friends with them, I drank with them till the end.

Social and cultural norms of socialising like drinking alcohol are a key component of building relationships in South Korea (Harkness 2013). In the university environment these often take the form of the 'membership training' that Mingyu refers to here. Her efforts to improve her tolerance for alcohol in advance of

South Korea in order to smooth these encounters across boundaries shows how initiative can also involve alterations in one's own body, in aligning differences in order to facilitate meetings and build connections with others. Other examples of bodily modification included adopting Korean fashion styles, learning colloquial-isms or being more familiar with Korean popular culture and play, each of which made the difference between Korean and international students seem more manageable in encounters; they provided possibilities to move beyond the con-tainment that sometimes characterised campus lives in a way that reaches out into wider urban encounters through the more diverse social assemblage that individuals become part of. While the kinds of initiative here clearly offer scope to improve relationships it is not something that all students could undertake. Indeed, there was considerable unevenness in the encounters that students had outside the campus, much of which was determined not only by initiative but also by the reception of difference in the city.

5.4.2   Uneven encounters

> Close friends? About two or three and out of the three two of them are Koreans right because in Korea it is not hard to make friends with [Koreans] because they are very very very friendly especially if you know the language and you know the culture it is like you get to blend instantly right so two of them are Koreans and I have lunch with them and dinner with them even on weekends if I'm free I'll just go hang out with them. (Kelvin, Malaysia, Male, Student KU)

There were a wide variety of accounts offered by international students on the kinds of encounters that they had in their everyday lives and the kinds of relationships that could be built with Korean students in particular, but also other Koreans in Seoul. A few, like Kelvin and students identified in Section 5.4.1, viewed positive outcomes in encounters as the result of the gen-uine openness of Koreans as well as their own familiarity with language and culture. In general, those students who had better language abilities also had better relationships with Korean students but because many students also studied in EMI courses advanced language ability was not widespread. Students with less language ability often suggested that they encountered blockages in encounters: 'we can't really understand each other' (Jiahui, China, Male, Student KU), 'I started like, to sense fear from students, like trying to speak English' (Arturo, Mexico, Male, Student SNU), '[language] is a restraint, it's a barrier' (Priya, Sri Lanka, Female, Student SNU), 'it is not as easy to be as close to them' (Shuchun, China, Female, Student SNU). These misalignments then reflect the particular configurations of globalising higher education in South Korea and the manner in which it articulates into the daily lives of stu-dents in Seoul. Put simply, the possibility of studying without Korean language opens access to a wider range of students but it also limits potential encounters with Korean students who are not familiar with or confident in English, it plays a role in configuring the character of student migration in Seoul.

There were also other blockages in relationships even where language was less of a concern. For the Muslim Malaysian students in this research that included the misalignment of Korean and Malay socio-cultural norms around pork and alcohol consumption. Hazmin (Malaysia, Male, Student KU), for example, described his Korean as being at a 'high level' but identified the limits of socialising as a key blockage in building relationships:

> Mostly I think the culture and maybe cause like maybe wrong, different, different, you know, different way of life because mostly we Malaysians most of them are like Muslim and they don't drink and you know that Koreans love to drink so you know like when they spend time with their friends they are, they want to drink so it's a little bit difficult.

Differences around socialising practices served as a key limit for many international students, particularly Muslim students but also those who were less familiar with alcohol consumption. While some, like Mingyu (China, Female, Student SNU), prepared actively for this possibility before arrival, participants unwilling to do so are then effectively excluded from social opportunities for meeting with and building relationships outside of the classroom. Held in contrast to Kelvin, who was Malaysian but of Chinese ethnicity and non-Muslim, Hazmin's account sheds light on some of the complex configurations at play in the encounters international students have with Korean counterparts – language, cultural familiarity and also socialising practices – that can reduce possibilities even for those with initiative.

More strikingly however, students also described in a range of ways the manner in which their very embodied presence was received differently in daily life in Seoul and how this influenced the character of encounters with others. For those who were less visibly different, like Korean-Chinese students for example, their daily lives were not interrupted by their own status as foreigners on a regular basis. Jeana (Mongolia, Female, Student SNU) captured this sense of ease:

> I more think that Mongolian. Koreans think that the international meaning is that America, Europe, like that. They never look at me like *weiguk saram* [foreign person]. Before I talk, they don't recognise me [as] the foreigner. Because the Koreans think that Mongolia and Korea is the same blood type, so that Mongolian is not foreigner! They say it like that! We are the same origin....When I go out and say I'm studying in SNU everybody says wow, you're smart!

Not being recognised as a foreigner plays an important role here in smoothing mobility through the city, a point that emerges in different ways in the next two chapters in relation to English teachers and across all migrants. When it is aligned with both assertions of ethnic affinity and the status accrued to students at elite universities as in the case of Jeana, these embodied characteristics can engender positive encounters and a sense of belonging that is not necessarily achievable for others. Indeed, while Jeana suggests her Korean-like appearance is unquestioned until she opens her mouth, other participants recounted the ways in which language disrupted their ancestral sense of connection to Korea:

> I was taking the subway with a friend. My friend and I were conversing in Chinese and we didn't do anything to this person at all, because we didn't know him. So, we were chatting with one another, with absolutely nothing to do with him.…He started speaking vulgarities, indirectly saying it to the two of us. He was directly hurling vulgarities at China and indirectly at us. So I felt that I couldn't understand why he had to act that way we didn't even offend him.…At that time, they were two guys, and we were two females. I just felt that it would not be good if we were to confront and quarrel with them at that time, so we just let it pass. But I felt very upset on the inside. (Huizhong, China, Student SNU)

Encountering racist demarcations, then, is not simply a result of being recognised as visibly different but also emerges through language use. For Huizhong this encounter and other occasions when 'Koreans generally treat me as a Chinese national' strengthened her sense of connection to China but it also undermined the ancestral links that had at least partly shaped her migration to Seoul in the first place; rather than identifying affinity, these encounters undermined her own sense of confidence about being in the city.

While only those who identified themselves as ethnically or ancestrally connected to Korea expressed this kind of tension between desired belonging and rejection in encounters, other more visibly different students reported uneven encounters in life on campus and in the city. In several cases, students reported that those with a Western appearance were received more favourably as *international* students, rather than regional students whose differences appeared to be less desirable:

> Well for the Koreans that I met most of them, most of them prefer to be friends with the Western people compared to the Asian people…because they are from the because they are from the developed country and they are from the like ermm much more big country something like that. (Masaki, Japanese, Male, Student KU)

In some instances, these hierarchies of desirability aligned not only with the ethnic and national differences of students but also the way they were placed alongside other migrants, such as those who come to South Korea for work. Two examples of misrecognition as a worker and student provide a sense of how this is experienced:

> In Seoul it was very very difficult, very difficult because you know I'm from Malaysia and then I have, I have brown skin and then I cannot speak Korean very well during

that time. ... So, it's very, very difficult and then people start looking at me from head to toe. [They] look at me and then when I start you know try to stop people say excuse me or something people will start looking at me and like woah is he dangerous or something like that. [When I went shopping] the person like just kind of like stop me and said don't touch it if you don't have money. ... I cannot communicate and then when they start talking I like cannot understand what they sort of like call me that you from Indonesia or you from Thai you work here in factory or something and then they they start treating you like like a labourer something like that. (Rayzal, Malaysia, Male, Student KU)

I was in a bus, other passenger asked me, are you working or studying? Are you studying? It feels like they give more respect. No, working. Oh what? If I knew, I would say that I am a student and show my student ID. (Edi, Indonesia, Male, EPS worker)

The notion of the desirable international student, the kinetic subject moving smoothly through campus and urban spaces is completely disrupted here by the racialisation of some bodies as associated with other derided migrant categories. In this encounter, Rayzal finds his body reconfigured as a working body that is denigrated rather than his own perception of himself as a high-achieving government scholarship student at an elite university. By contrast, Edi learns he is positioned not only as different from Koreans but also relative to other migrant categories such as international students. Such examples reveal the way that the seemingly different migrant lives of students and workers can be configured in relation to wider discourses around race and the value of migrants. Céline (France/USA, Female, Student SNU) provided another example that revealed an alignment with English teachers that she also wished to reject:

They'll stare at foreigners, and they have all these preconceived ideas! Like oh hi, you know, you're Westerner, you must be an English teacher! And actually, I was with my friend on the subway who's American today, and neither of us are that! Like, my friend is married to a Korean, and so she's a housewife here, and I'm a student! ... I feel like people – when everything is homogenous, they start to think they know everything about everyone. And ... it's not true. (Céline, France/USA, Female, Student SNU)

In both these instances, embodied difference is aligned with language use and the socio-political position of migrants. Put bluntly, the experience of many students was that if they were seen to be from South or Southeast Asia they were thought of as labour migrants and for white international students they were perceived to be English teachers. Such perceptions align with earlier reflections on the way that migration regimes work through perceptions of ostensible national and ethnic differences and developmental levels. Alongside differences in language and the capacity and drive of students themselves to initiate encounters with Koreans and others in Seoul, these encounters reveal the uneven encounters that migrants have in the city, configured not only by their socio-legal position as more or less desirable subjects but also the manner that their embodiment matches with the discursive renderings of these subjectivities.

5.5   Conclusion

The articulation of desiring-migration with urban assemblages and the encounters that emerge in migration can be productively unpacked through analysis of the purportedly kinetic movements of desirable migrants like international students. Seoul and South Korea's aspiration to become a knowledge centre and a node in highly skilled migration has been a key component of the globalisation of higher education and the initiative to attract and retain international students at elite universities like KU and SNU. As this chapter has shown, substantial effort is invested by universities and the government in actively assembling these migration flows through the promotion of study opportunities but also the establishment of infrastructures of arrival and targeted spaces on campus that seek to smooth the flow of international students. These strategies constitute a territorialisation of migration into specific urban spaces, not only facilitating flows but serving to establish the privilege of this pattern of migration over other forms. The discrepancies generated in such initiatives are revealed particularly starkly in the comparison between the management of highly sought-after scholarship students and the compulsory orientation and control programmes established for EPS workers. Others, whether less desirable students or English teachers may be handled with more ambivalence but there too we can observe the active assembly of migrants' urban lives, in campuses, workplaces and housing but also more widely in the city. This is the production of discrepant experiences not only in migration but also in becoming part of urban assemblages.

There is also something particular about the accounts of international students offered here that draws attention to the unintended effects of managing migration and its outcomes. The territorialisation of international student migrations in universities in Seoul and their broader lives in the city discussed here demonstrate both instances of mobility but also significant channelling and containment. Students encounter campus spaces that are already striated by perceptions on the value of diversity and their place as international subjects who can enhance status but who also need to be managed in order to smooth their arrival and adaptation. The effect as the accounts of students demonstrate is a significant containment of international student lives on campus whereby students are channelled into particular courses, administrative spaces and social networks that reduce possibilities for interpenetration while also making mobility easier.

In this sense, the chapter draws attention to the complex intersections between the politico-institutional desires for enhancing circulation and the very desires that are claimed to be cultivated in international student mobility. There is most notably a tension between the drive to open university campuses to global flows and retain students as future potential human capital and the practices of managing these flows that seem to lead to containment that might stifle the desires of students and the potential encounters they can have. English language has been a particularly important component of this as well as international residence and administrative spaces. These new campus assemblages liberate international student mobility because they generate the imaginative potential of study in Seoul for a wider range of students while also containing their lives through a segmentation of classroom and campus spaces that also reaches into urban lives in Seoul. The *international* or even *foreign* embodiments of students are amplified as a result and the building of relationships with Korean students is experienced as particularly frustrating. Here, then, the challenges of assembling and inserting talent flows for the reconfiguration of university orientations is impeded by the differences that this desire for migration brings.

Lastly, this chapter shows that for international students as for other migrant groups, these territorialisations within the specific sites of emplacement in the city (campus, workplace, housing) also reach out into wider urban lives. Chapter 4 has already provided some sense of this in the manner that the very literal as well as metaphorical distance of the periphery works to constrain the urban lives of EPS workers and adds to their rendering as marginal and undesirable working subjects. Chapter 6 too will consider the way that English teacher lives in the city are shaped by their associations with whiteness but also the particular social, spatial and temporal configuration of their working lives. The accounts of student encounters in the city discussed at the end of this chapter provide another dimension of these configurations. Unimpeded by the temporal constraints of work life and embedded in the logics of globalising higher education, students might be presumed to have time, financial resources and desire to engage and encounter others and build meaningful relations with the city. While there are instances of encounters and the building of relationships the discussion here has shown it is unevenly experienced in ways that reveal the intersection of the containing logics of globalising higher education and wider discourses on the embodiment of desirable migranthood in the city. Students are clearly not received equally in the city and while some students find their bodies aligned with desirable components of globalising higher education and connections to a sense of being kin-like for Koreans, others are racialised as undesirable foreigners in ways that align them with other subjects like migrant workers and English teachers. These uneven experiences reveal the ongoing significance of codifications of ethnicity and nationality in the lives that students and others live in the city, as well as the way that the logics of the migration regime reach beyond its immediate effects on legal status into the possibilities for life and sociality in the city.

Endnotes

1   The other students without government or corporate scholarships included two exchange students (Japan and USA), a graduate student from the USA and a student from Mexico whose father had been transferred to Seoul by his company.
2   Myeongdong is a prominent shopping district in the centre of Seoul. It is made up of a dense grid of pedestrian malls and alleys with a vast array of designer and street fashion, food and technology. There are also several major department stores in the vicinity. It is well known for the large number of tourists and visitors who frequent it; retail staff are often proficient in Japanese, Mandarin and English.

Chapter Six
Negotiating Privilege and Precarity in Suburban Seoul

As a Westerner, you probably know as well, some people just stare at you, and you know, you don't know whether it's cos' they're like 'Oh my god, a Westerner!', or whether it's like 'Oh, god, a Westerner. What are they doing in my neighbourhood' that kinda thing, you know? It's really hard to know what they kind of mean when they look at you. I don't know. (Nadia, UK, Female, English teacher)

In Chapter 4 it was clear that visibility and invisibility contribute in significant ways to the lives of migrants in the city, shaping the extent to which migrants can participate in different urban spaces and at times serving as a basis for a politics of recognition. Visibility and invisibility also have a significant role to play in the character of encounters that migrants have and in the ways in which migrant subjectivities are assembled in relation to the perceptions of majority populations and the way migrants perceive themselves in society. Immigration controls established in the migration regime, then, form only one component shaping the discrepant experiences of life and work that migrants have. We also need to consider the ways in which migrant subjectivities and experiences are constituted through particular racialised, gendered and aged assessments of value and possibility (Anderson 2009). Nadia, like many other English teachers, EPS workers and international students in this research was struck by her visibility in Seoul and the extent to which her body attracted attention. While varying levels of visibility were shared amongst many participants, however, it was also clear that there were considerable differences between the ways in which different migrants were received and responded to in encounters, differences that manifest both at

Global Asian City: Migration, Desire and the Politics of Encounter in 21st Century Seoul, First Edition. Francis L. Collins.
© 2018 John Wiley & Sons Ltd. Published 2018 by John Wiley & Sons Ltd.

the level of different visa status – worker, teacher, student – but also in other con-figurations. For many teachers, visibility was first and foremost associated with whiteness and its privileged status in relation to foreignness in South Korea. Dean (Australia, Male, English teacher) provided an indicative account:

> I have found that with Koreans, if a Westerner takes some interest in their culture, they are typically pretty impressed by it. So, I have always been treated really well even though I can't speak Korean at all. There are ones there that can speak English and that sort of thing, so I have had mostly positive experiences from those places. Treated like sort of a token white guy, and they are sort of keen on it because that gives them the feeling their centre is a lot more international than other places.

While being the 'token white guy' in the way described by Dean here is hardly an example of deep relationships it clearly reflects a relatively privileged position in everyday life that relies on the association of many but not all English teachers with whiteness. As I have already noted in Chapter 4, teachers tend to live in rather ordinary suburban neighbourhoods, characterised by significant residential development, schools, commercial and retail areas and established infrastructure. These spaces provide a multitude of opportunities for contact with local populations that enhance the experience and sense of inclusion that teachers have. As relatively rare 'westerners' in Seoul's neighbourhoods, many teachers commented on the extent to which people were curious about their lives. Many reported instances of residents seeking to practice English with them, of elderly women looking out for their wellbeing or of shopkeepers treating them to free goods. Others reported how even simple efforts to learn Korean could engender positive responses from their neighbours. Such practices reflect hospitality that while certainly positive are also tied up with the desirability of the West and westerners rather than these individuals themselves. They contrast significantly with experiences of EPS workers, such as Nonoy (Philippines, Male, EPS worker) below, as well as other migrants whose embodied experiences did not align with whiteness or other privileged forms:

> If you're Korean, you're nothing but Korean. They don't want to, to, to mix with migrants. … I experienced firsthand all of the, all the Korean places I'd been to since, I got a taste of it, I faced it during my time there, all kinds of attitudes. All of it, people swearing at me, insulting me, yelling at me, you name it. Because, like I said, the workers there were traditional Koreans. It's like they couldn't accept that there were foreigners working there, … it was all right in the office but in the production area, it was like, they couldn't accept it, especially if you're dark-skinned, then they really abuse you. (Nonoy, Philippines, Male, EPS worker)

Race, class and migrant status play out in multiple ways in this extract and comparisons that can be made with Dean's account above. It is clear to start with that the embodied differences of migrants influence their encounters with

Koreans, with westerners who associate with whiteness experiencing positive if tokenistic embrace while 'dark-skinned' workers can be openly derided. The prevalence of these racialised framings of migrants in Seoul was also brought to the fore by international students who narrated accounts of being misrecognised as either English teachers or EPS workers. Recall in Chapter 5 that Céline (France/USA, Female, SNU Student) described how people she met presumed if 'you're a westerner, you must be an English teacher'; and Rayzal (Malaysia, Male, KU Student) described being told 'you work here in factory or something and then they start treating you like *no dong* [labour]'. Embodiment matters in shaping migrant encounters in ways that exceed the formal statuses established in the migration regime and in ways that mean that migrant subjectivities are socially constructed beyond the control of migrants themselves. International students can be rendered as teacher or labourer in ways that they do not desire unless they can demonstrate their association with prestigious universities.

Similarly, teachers who were not seen as white experienced quite different encounters. Alex (South Africa, Male, English teacher) described feeling like a 'travelling freak show' and how when 'I sit in the bus and there's one seat left next to me and everyone like rushes in and they see the seat and they're like uhhhh and they see me and they're like oh fine I'll stand and so it's like, okay fine dude, it's cool, I'll bite you later!' Charlotte (USA, Female, English teacher), a Korean adoptee who was raised in a 'Caucasian household' expressed insight into these configurations:

> I also feel like it's a privilege as well to be able to have invisibility. [I have] two friends who are I guess, they are identified as Black American. When I go out with them, I'm just shocked by the amount of stares, pointing and reactions I get. It's like they just distort whatever area they're in, everyone has to react back in some way or they don't really feel comfortable. I can't imagine the psychological toll it makes.

Charlotte's account points to the different role that racialised understandings of foreign bodies play in the encounters that teachers have in place. There were relatively few teachers who identified as not being white in this research, but amongst them those who identified as Asian like Charlotte found that they were relatively invisible in their daily lives while as Charlotte indicates here for black teachers the encounter with local residents is shaped by quite different racialised contours. Here the privilege associated with either being recognised as white and associated with desirable foreignness or being able to be 'invisible' and blend in is laid bare.

This chapter takes these insights that emerge in embodied encounters as a starting point to consider the ways in which migrants negotiate their often-contradictory position in Seoul. The focus here is primarily on English teachers, who occupy a peculiar position in Seoul and South Korea as seemingly desirable foreigners, established in part through racialised readings of their embodiment and a status as teachers and sojourners in the migration regime rather than workers. Unlike the narratives of EPS workers prioritised in Chapter 4, English teachers have considerable freedom to direct their own mobilities, to enter and leave South Korea as they please, to renew their visas indefinitely, and in some circumstances to transition to long-term residence. These freedoms rest in no small part on the relationship maintained between whiteness, life stage and desirability in the migration regime and on the framing of English teaching as a short-term professional rather than labouring occupation. English teachers have an ability to move into an occupational niche that is exclusively available to them because of their privileged national and educational status. Yet, as this chapter demonstrates, the social milieu of English teachers is also produced in ways that are character- ised by considerable transience and limited attachment to place. These condi- tions are produced in the migration regime but also in employment relations and the particular social practices of English teachers. Many are in Seoul for relatively short periods of one or two years, but even for those who remain longer their nar- ratives foreground feelings of floating and drifting between contracts, of becoming paradoxically immobile in their mobile lives.

By focusing on this peculiar intersection of privilege and precarity in English teacher lives the chapter offers important insights into the ways in which migrant subjectivities take shape through the articulation of desiring-migration, the conditions and driving impulses of migration and the configuration of particular migrant lives in workplaces, daily life and societal discourses. I high- light in particular the temporal dimensions of migration as this emerges in the constant circulation of teachers and the churning of visas and address the way that this departs from EPS workers and international students. As the chapter shows it is exactly the inability to reconcile their current position with desirable or predictable futures that reveals precarity in teacher lives; unlike EPS workers they do not face absolute time limits or a subversive move to undocumented status and unlike international students they are not positioned on a future-ori- ented trajectory of self-development through education and graduation. There is, then, both privilege and precarity in the formation of this migrant subjec- tivity and the expectations associated with it. At the same time, the chapter also reveals how some English teachers exceed these expectations through the reconstitution of subjectivity that occurs in the building of more permanent social relations and community connections, the desiring of other possibilities for migration and urban life.

6.1   Privilege and Precarity in Migrant Subjectivities

Migration is regularly represented as a disruptive or deterritorialising force. Often this assertion is made in relation to the effects of migration on different geogra- phies – world, nation, region, city or community, as we have seen in the previous chapters – but we can equally observe that migration is disruptive in terms of the discursive and material formations that support the establishment and maintenance of coherent subjectivities. In migration, people become rendered as migrants through migration regimes in ways that are beyond their own control (Anderson 2009). As is clear in the examples already used in this chapter, this migrant subjectivity is not universal but rather is expressed in a wide variety of ways that operate in relation to racialised, classed, gendered and age based interpretations of embodiment. 'The migrant', as Anderson (2009: 408–409) documents, 'is not simply a legal construction, and there are multiple ways and scales by which this figure is imagined, defined and represented'. This includes the categories established in migration regimes but also racialised, gendered and aged social constructions, individual circumstances and desires and social and employment relations. In terms of this research, for example, it is observable that the migrant worker becomes codified as male, working class and South or Southeast Asian, the international student is associated with youthfulness and globally mobile middle-class cosmopolitanism and the English teacher is situated as young, white and ideally North American. These migrant subjectivities are contested fields, however; they involve the social construction of different forms of migration and labour but can also operate as part of migrant desires to reconfigure or transform subjectivity (Cheng 2011).

The migration of English teachers from western Anglophone countries to South Korea and other parts of Asia is often framed within a discourse that emphasises youth, freedom of mobility and whiteness (Lan 2011). Within this discourse, English teachers are constituted as relatively privileged and exceptional subjects of migration systems. They are privileged because their access to work visas is relatively unconstrained, they often have the legal capacity to renew visas and remain for long periods of time and because their association with the idea of 'the West' and use of English language carries cultural capital that eases their mobility across culturally diverse territories (Phan 2016). This privileged status also means that English teachers are seen as exceptional to other forms of migration and particularly the kinds of labour migration discussed in Chapter 4 (Lundström 2014). Indeed, despite the work-based character of their migration English teachers are very clearly *not* constructed as migrant workers in the ways in which their migration is managed and in the relationships produced in the workplace and urban society. Moreover, in policy and academic accounts of migration in South Korea English teachers are almost completely absent as individuals who might be understood as migrants. Instead their migration is socially constructed as a kind of travel–work experience that is short-term, connected to particular life stage periods and embedded in cultural rather than economic drivers of migration (Collins 2014b).

The framing of English teachers as privileged and exceptional subjects of migration in Asia is closely connected to associations with whiteness that condition their socio-cultural positions and work lives. Indeed, as Lundström

(2014) points out in *White Migrations*, 'the migrant' has been a discursive marker that is almost always used in relation to non-whiteness and being non-western (see also Nail 2015). In contrast:

> 'white migrants' can inhabit the world as part of a global enterprise, tourists, expatriates, guests, developmental aid workers, and so on, representing humanity, whose presence remains undisputed or who are able to use their white ethnicity as a form of 'symbolic ethnicity'. (Lundström 2014: 2)

The contrast with discourses around migrant workers as problematic and in need of saving in South Korea specifically (Kim, N. 2012; Kim, S. 2012) and internationally (Anderson 2013; Bauder 2016) could not be starker. Even international students can be touted as both a talented current or future elite as well as 'backdoor migrants', 'cash cows' and 'victims' (Robertson 2011). By contrast, whiteness is strongly associated with privileged forms of migration both in popular discourse as well as in the recent growth of scholarship on expatriates or mobile professionals that emphasises their difference from other kinds of migrants. Fechter and Walsh's (2010) overview of the field, for example, emphasises the need to distinguish what they call 'mobile professionals' from other kinds of South–North and South–South migration. In this context, privilege is associated with undertaking lifestyle migration as a 'choice' rather than an imperative of economic situations (Botterill 2016).

There is in my view an alignment between these scholarly and popular understandings of white migrations as privileged and less economically driven and the ways in which particular individuals are then viewed in migration contexts. As I have already discussed in Chapter 3, the recruitment of English teachers signals the importance of being identified as a native speaker and the working opportunities available to these individuals being viewed as a white American is idealised by employers, students and parents (Ahn 2014). The notion of the imagined West is critical here as are the various hierarchies that cut across this geography (Phan 2016). Other markers of difference clearly intersect with whiteness and nationality in this context too. Normative conceptualisations of life course are particularly important as English teachers are understood as being relatively young recent graduates who do not envisage long-term futures in South Korea. Gender is also significant: while men or women can be more desirable as English teachers in different educational settings, it is clear that moral panics around English teachers have revolved principally around the representation of white men, their social behaviour and their relationships with Korean women and children. More widely and paralleling wider imaginations of young travellers, it is a particular configuration of English teacher-hood that is desired with preferences around nationality, age, ethnicity and qualification that influence the location, remuneration and availability of jobs (Collins & Shubin 2015).

At the same time, however, migrant subjectivities are also constituted through the desires or social forces that generate migration and the conditions through which migrants become part of places. In this respect, following Cheng (2011), we need to examine not only the ways in which migrants are differentially objects of desire and/or derision but also their role as subjects of desire, who bring into migration their own imaginings, expectations and capacities to alter their position as migrants. The preceding chapters have already shown us that migrant workers, for example, can rework marginalised positions through mobile communing, subversion and tactics of recognition and international students' narratives also speak to the capacity of some migrants to reconfigure relationships with local populations.

The lives of English teachers offer further insight into the tensions that exist between the legally and socially constructed status of migrants and the individual and shared conditions of people's migration. Many English teachers in this and other research (Lan 2011) construct narratives of themselves as travellers and adventurers, seeking overseas experience as self-development in ways that echo normative understandings of mobile professionals and young western travellers (Fechter & Walsh 2010). Yet, at the same time the drivers for migration, such as that described by Nadia in Chapter 3 situate migration also with the expansion of higher education, qualification inflation and underemployment at home, and the burden of student and other debts. Lan (2011: 1676) came to a similar conclusion in her research in Taiwan, where she argued that:

> Under the rosy images of global trekkers and cultural adventurers, however, lies the dark current of economic migration and marginal employment.... Some faced poor job prospects in the local labour market upon graduation and looked into alternative possibilities overseas. Some had worked in their home countries for a few years but found themselves trapped in low-paying, entry-level positions with little career promise.

Understanding this form of migration in relation to privilege then requires considerable nuance and raises the important question of connecting privilege to ideas of marginalisation and precarity that characterise English teacher and other forms of migration. In this chapter, I seek to move beyond the typical view of privilege and precarity as situated in a dichotomous relationship (Botterill 2016) and the separation of English teacher and other white migrations from analysis as forms of migration. I draw attention to the manner in which English teachers negotiate their own precarity through expressions of privilege that can in themselves shroud feelings of vulnerability. In doing so I take a view of precarity that is not constrained to specific categorical conditions but rather a broader 'unbound' notion of precarity as 'a condition of vulnerability relative to contingency and the inability to predict' (Ettlinger 2007: 320). In this respect precarity is situated not in opposition to privilege but rather to the certainty and security that were expressed in modernist political projects (Ettlinger 2007; Tsing 2015). Migration can be an important part of these configurations because it is often a response to uncertainty at home and yet the process of moving necessarily involves uprooting and disruption that has an embodied cost on those involved (Nail 2015).

Although it is possible to identify precarity as an important generator of migration and as something that is reworked through migratory processes, my focus on the precarious–privilege of English teachers draws particular attention to the articulation of these conditions in the spaces of everyday life (Hall, King & Finlay 2016). If, as Tsing (2015: 29) asserts, precarity revolves around 'our vulnerability to others' then it is also necessarily embedded in everyday life as the spaces in which we interact with others and think and feel about it (Ettlinger 2007). Indeed, it is in everyday life that notions of (un)certainty and (in)security about present and future lives are felt in our daily activities, planning for the future, imagining other possibilities and interacting with others who seem to share or depart from our current position. Uncovering the precarity that is shrouded in articulations of privilege and freedom then demands attention to the less certain dimensions of migrant lives. For migrants, precarity emerges in uncertainty about the future, about the terms of contracts or the ability to maintain employment, build or nurture a family, establish residence rights or citizenship, remit money or fulfil promises (Hall, King & Finlay 2016). Migration management regimes then are centrally implicated in generating precarious lives (Lewis et al. 2015) but they do so through connections with individual migrant trajectories through places and in concert with others in these places.

6.2   'I teach to live, I don't live to teach' (Charlotte, USA, Female, English teacher)

English teacher migration was rarely articulated through desires associated with specific forms of work and occupation but rather through more loosely expressed ideas about opportunity, escape and transformation. The occupation of English teaching or even education more generally only figured in the biographies of a very small number of interviewees in this research. Rather, many had taken opportunities in South Korea not to teach or develop a career but rather, as Charlotte puts it, *in order to live*. Their migration often revolved around underemployment, debt and finding ways to travel when they had little access to finance.

There are important similarities here in the narratives of both migrant workers and international students, which in both cases included many individuals who saw migration as taking a chance and exploring the world. Dewi (Indonesia, Female, EPS worker), for example, described her excitement before coming to Seoul: 'I was excited every time I watched Korean movie, I imagined what it would be like when I go to the country'; and Yuming (China, Female, KU Student) described a frustration in not getting into a good university and the interest in trying something else through migration, 'ever since I started university in China, I had been having the desire of going overseas.' While other individuals crafted more goal-directed narratives around earning income or gaining credentials these instances remind us that migration also involves people being taken along by the possibility of being elsewhere, by desires to become or avoid something through moving in the world.

The structure of English teacher recruitment articulates in important ways with the pre-migration situation of teachers. Participants spoke of finding internet advertisements for teaching jobs that led into an expedited process of recruitment, departure and orientation. Many described a long held interest in travel but very few had been planning or making preparations for this in the near future or had considered South Korea as a desirable destination. Migration, then, was not described as part of linear plans for life but rather as a kind of seize-the-moment opportunity that must be taken before it is lost. Accordingly, many teachers found themselves not only suddenly in another country they knew little or nothing about but also in a job they had no training for nor any necessary interest in doing.

Jarod's (USA, Male, English teacher) account was relatively indicative of this fast paced transition from searching for jobs unsuccessfully at home to suddenly finding himself teaching English in a classroom. Unlike some other participants, Jarod had some knowledge of Korea because his father had worked for the US Government there in the 1970s. Like other teachers, however, he used a recruiter who very quickly identified a job for him and arranged tickets. Within a month he was flying, from Seattle via Tokyo to Incheon. He was picked up by his employer at the airport, taken for a large dinner and then the next morning he was asked to start work:

> First impression of Korea you know, great airport, very modern. We went out to dinner that night. I would say that I was, if anything was overwhelming it wasn't cultural, it was really just the job. It was like 'Boom' like next day start teaching.... I could imagine some people might want to decompress but there was none of that.

The rapid transition from arrival to teaching reflects the high demand that has existed for English teachers at certain times, particularly in private academies that cater for children where most teachers work. Those who took contracts with public schools through programmes like Gyeonggi English Program in Korea (GEPIK) were ordinarily provided two weeks orientation and preparation before beginning their work.

Teaching English is at once the main reason for English teachers' presence in South Korea and yet at the same time is also expressed as a site of seeming dislocation from everyday norms. While work hours vary, English teachers generally tend to work 25–30 hours a week. These work hours contrast significantly both from perceived norms around career occupations in the countries teachers come from as well as from the working hours of their Korean colleagues (to say nothing of EPS workers discussed in Chapter 4). The result was a sense for many that their job just didn't make sense in the temporal terms they were familiar with:

> I've got teachers, the working hour is, hysterical funny. Eight hours a week, other people work 45 and you work from nine to five, you work from half past eight, you stay until half past four but you don't work half of the time. It's like a holiday actually. (Chloe, South Africa, Female, English teacher)

Many teachers expressed this level of positive surprise at the time that they were expected to work in schools or academies. In contrast to notions of the ordinary 9–5 day that characterises middle-class expectations and norms (if not always realities) these work hours seemed filled with freedom.

> Well the class types and the students change every day. I will always be at work, the times I am expected to be at work if I am needed is from 7–11 am and 5:30–9:30 pm and if my first class is at 8 am, I can show up just before 8 and it would be fine. If my last class ends at 7 pm, I am done at 7, there is no need to be at work if you are not working. (Craig, USA, Male, English teacher)

Craig's position was relatively unique in that he worked for an academy where adult students scheduled their own appointments and classes. Even for teachers who had fixed hours, however, such as those at public schools, they were rarely expected to work through the day but rather found themselves with considerable periods of time where they had to find other activities: 'don't work half of the time', need to 'fill time' in-between classes with social media or other activities.

> I would teach, depending on the schedule but usually I would teach three classes in the morning. I'd be done at 10. And then I would go home, take a nap. I'd sleep till about two. I would wake up, probably do some reading, read a book or play some video games, something to keep my mind off… whatever. Then I would get ready for work again around six, I guess and class would begin at seven. Then again I would teach three more classes. So three in the morning, three in the evening. I'd be done at 10 and at 10 I would head over to the internet cafe.… Just basically, go to school, nap. Go to school, hit the cafe and sleep. (Curtis, Canada, Male, English teacher)

Like many other teachers, Curtis described his working hours as ideal because of the free time that they gave him; his and other accounts also point to the ways in which English teachers are very much exceptional working subjects in South Korea that can generate forms of dislocation in their lives and at times precarity for those who desire greater stability. The notion of working less than six hours a day and earning enough to enjoy a social life was presented as a positive dimension for many teachers. It also reflected the ways in which teachers are held, and hold themselves, at a distance from involvement in work life:

I also think that English teachers tend to be very bad at this because their jobs dictate that they spend the majority of their time not integrating – they're almost a little window into the language and the culture of the English-speaking world, and that's what you're paid for, that's what you're here for. ... There is an element of *an English teacher is a machine that can speak English and can teach us cultural aspects of the English-speaking world*. (William, UK, Male, English teacher, emphasis added)

As a 'machine that can speak English' many teachers' work lives departed from temporal and social norms in South Korea while at the same time they were disconnected from norms in the countries they came from. These disconnections became particularly apparent in workplace dynamics, as Margaret (UK, Female, English teacher) indicates:

We all worked in the same office and because we shared classes we had to talk about those classes. I think we felt that a lot of the time the Koreans would talk amongst themselves in Korean and we felt a bit left out cos' we couldn't understand everything. ... Miscommunication sometimes as well perhaps, but generally we all got on really well. I think the problems arose when they would have a conversation about something in the teachers' room and we weren't involved because they were speaking Korean perhaps, but maybe they thought they had established something and they decided on something, but we wouldn't get told about it. I think that was the main problem – we'd find out about things at the last minute.

Many teachers including Margaret often gave positive accounts of their Korean colleagues but they also noted the extent to which their role in the workplace was minimised to a functional speaker of conversational English rather than a professional colleague who might be part of institutional operations. Their position as migrants in Seoul and South Korea is produced in such arrangements, where labour relations align with expectations of teachers as only short-term sojourners and not as either migrant workers or potential long-term colleagues or residents.

These accounts point to some of the arrangements that teachers encounter in their migration to Seoul and a starting point to consider the ways in which different forms of desiring-migration influence how individuals negotiate their position as migrants. In the case of English teachers, for participants who envisaged short-term stays in South Korea the idea of being 'a machine to teach English' appeared relatively unproblematic, aligning with preconceptions of temporary work as a means to enjoy time doing other things. Colin (South Africa, Male, English teacher,), for example, lamented about people who 'are definitely out here with a purpose [because] that kind of for me, restricts the fun you can have, because it is not like a spur of the moment, let's go party; ... this is supposed to be fun, we are young'. Other teachers, especially those who had remained in Seoul or South Korea for some time, saw a need to separate themselves from the highly transient dimensions of English teacher lives and to craft other forms of attachment and belonging in the city. These points are taken up in greater detail in Section 6.4 of this chapter. Section 6.3 considers first the impact of turnover and transience.

6.3   Turnover and Transience

A key component of the ways in which different forms of desiring-migration become produced around particular migrant subjectivities and articulated into urban assemblages relates to the temporal patterns of migration, of migrant lives and of the spaces that they move through. Time, in this respect, should not only be read as an objective measure of different moments in migration but rather needs to also be examined as a constitutive feature of the subjectivities and encounters of migrants (Anderson 2009; Collins & Shubin 2015; Griffiths 2014). This is evident in the 'temporal constraints' that set time limits on visas or employment contracts (Robertson 2014) and also the rhythms that condition the daily lives of migrants (Marcu 2017).

The EPS migration regime, for example, sets strict time limits on the duration that migrants can legally spend in South Korea at four years and ten months, and individuals who seek to exceed this become marked as undocumented and illegal. The daily lives of workers are also characterised by a quite intensive focus on work activities. Surveys show that 50% of E9 visa holders work more than 50 hours per week and 30% work more than 60 hours per week (Korea Immigration Service 2017). Many participants in this research reported working six or even seven days each week from early in the morning and sometimes until very late at night as a result of overtime opportunities or split shift requirements. As Chapter 4 demonstrated, these temporal rhythms reproduce migrants' position in the urban periphery because they reduce opportunities for social contact outside worksites. Put another way, they contribute to the constitution of a migrant worker subjectivity, framed in legal terms that then get enhanced through particular employment relations and their impacts on the wider urban lives of migrants.

By contrast, the temporal constraints and rhythms encountered by international students are maintained in relation to the institutional framework of universities. The length of visas is determined by course of study and the ability to pay fees, and the rhythm of migration and daily life emerges in relation to the calendrical patterns of semester and daily class, study and extra-curricular activities. Student lives and migration are accordingly oriented around the temporal patterns of the institution and the requirement for presence at certain times within campus spaces. Their accounts, as revealed in the Chapter 5, are unsurprisingly characterised by much greater sense of freedom in terms of time or even a sense of having too much time. The future then is crafted as an inherent component of international student subjectivities, not only in terms of their progression through educational spaces and the opportunities available to explore and learn in the city but also because they are positioned as potential future residents and citizens. In contrast, to both migrant workers and English teachers they are conceived as individuals who could be transformed into future human capital to address the challenges of demographic renewal and economic advancement in South Korea (Shin & Choi 2015).

These brief interludes into the timing of worker and student lives give some sense of how diverse temporalities feed into the conditions of migrant lives and contribute to discrepant experiences and subjectivities in the city. The timing of English teacher lives in Seoul intersects in curious ways with both students and workers. On the one hand teacher lives are oriented around work-related activities like EPS workers and they also face visa-related temporal constraints, but on the other hand the narratives of teachers also foreground a sense of having too much time or being uncertain about the progression of time. These patterns are apparent in the discussion of workplaces in Section 6.1 that suggests many teachers are viewed, and view themselves, as temporary conversational additions to educational institutions. Their everyday work-times and future progressions are structured around this orientation. They also align with expressions of being 'permanently temporary' (Collins 2012; Rajkumar et al. 2012) that is captured neatly here by Chris (Canada, Male, Teacher):

> So yeah, I arrived. I guess I still celebrate it. My Korean birthday is November 2nd. So, every November 2nd is like 'Yeah! Another year!' (Chris, Canada, Male, English teacher)

Chris first came to South Korea at the end of 1996 just before the 1997 Asian financial crisis; he had remained for 13 years at the time of the interview. He came seeking an opportunity to pay back student debt but not long after arrival found himself in even more difficulty as the collapse of the Korean Won meant that he was unable to pay his debts in Canada and he was in effect required to stay longer than planned in order to pay back overdue charges. Over time, Chris kept on signing the 12-month contract that is standard for teachers; he moved workplaces, gained more experience and started to become relatively well established in the community of long-term teachers. After more than a decade in South Korea, though, Chris still found it hard to envisage a desirable future elsewhere. He now works at a university that has 'not got the greatest reputation' where the pay wasn't great but where he considers he has 'the best job in my life'. Despite this job, his life remains tenuous, oriented around a twelve-month contract cycle that inevitably returns each year on November 2nd, his 'Korean birthday'.

This example speaks to the wider ordering of English teacher lives around temporalities generated in the migration regime that governs teacher mobility. The E2 visa can be issued for up to two years but in most cases is issued for twelve months and must be sponsored by an employer involved in English language training. While the renewal itself is not necessarily an arduous process and can be an opportunity for individuals to move workplaces and/or cities, it also reinforces temporariness in teacher lives by constraining their life choices in South Korea to this very particular occupational niche. These 'temporal constraints' (Robertson 2014) operate around the socially constructed notion of who an English teacher is and what their duration of residence in South Korea should be, they are expected to be relatively young university graduates seeking short-term employment overseas not migrants seeking long-term careers and future citizenship. As in the case of the EPS regime, the temporal constraints of the E2 visa operate as a governmental technology that limits or controls how migrants live in South Korea by shaping who they can be employed by as well as what sorts of rights of residence and access to social resources they have.

The lives of English teachers are structured around these temporal constraints. Indeed, unlike migrants moving through the EPS regime, which involves visas with longer durations, the annual contracts and visas of teachers articulate with high turnover and a sense of transience. There is considerable 'churn' in English teacher populations in Seoul, as new teachers arrive to fill the spaces opened by those departing and, in the more precarious private academy business, schools close down or open in new locations. As Craig (Canada, Male, English teacher) explains, this transience has a significant effect on how many teachers relate to Seoul and their place in it:

> There is not this super hard-core group of people that is always there, we are all teachers on a one-year contract, the vast majority anyway. The average person comes to Korea, last I heard was about 10 months, was the average teacher's stay in Korea, from a person who just got here to a person who has been here for 10 years, the average person is here for 10 months. So we are talking a fairly short period of time to really call a place home.

While the churn in teachers is particularly notable for generating transience in new arrivals and early departures, its effects are also apparent for longer-term teachers. Indeed, even those who have been in South Korea for several years see the annual contract signing as a significant temporal marker, a potential closure on their duration of residence. Roger (Canada, Male, English teacher) and Dean (Australia, Male, English teacher) who had been in South Korea for more than five years at the time of the interview put it this way:

> Each contract is a separate year for me. I'm not looking beyond the end of my contract. And actually, it is possible that some fantastic job came up somewhere else, I would take it. (Roger)
>
> Something else about the foreign teachers here is we never know how long we will be here. None of us would think about setting up permanent roots or anything as we are not sure where the next year contract will be renewed, we know we are only here temporarily so we try to do that less. (Dean)

Like Roger and Dean, many English teachers spoke of temporal horizons that were characterised by considerable uncertainty. In this sense, English teachers are not faced with the closure that exists in the strict temporal constraints of the EPS regime but neither are their migration pathways established on a relatively stable progression through stages as is the case for international students. The annual contract signing provides a level of employment prospects for the near future but does not offer certainty beyond that. Some longer-term teachers even spoke about needing to balance a desire for career progression with their desire to remain in, or at least to avoid leaving, South Korea. Many private academies, for instance, favour newer graduates who can be paid less and have less expectations in their jobs and as such the annual contract can generate uncertainty about the future for longer-term teachers. It is partly for this reason that many seek further qualifications to progress into the university sector where remuneration is not necessarily greater but where there is more certainty around contract renewal.

The turnover and transience in English teacher populations has considerable implications for the emergence of the kind of social connections that were described as 'mobile commons' (Papadopoulos & Tsianos 2013) in Chapter 4. Recall that these commons were generated amongst workers remaining in Seoul for long periods of time, or even amongst international students through the formal support of institutions and in relations between senior and junior cohorts. In such contexts accumulated experience can be shared or sold amongst contacts, within or across co-national boundaries, in ways that open up new possibilities for life in the city. For English teachers, the churn involved in migration undermines the manifestation of some key elements of these commons. Certainly, knowledge, information and 'tricks for survival' circulate widely, either through interpersonal relations or through internet based forums and social media that are both specific to South Korea or connect into more geographically extensive English teaching networks. At the same time, there is much less evidence of solidarity and mutual care amongst teachers, and even forms of sociability seem more constrained. As noted in Chapter 4, for example, movements like the Association of Teachers of English in Korea have emerged in response to specific events and around motivated individuals but decline as attentions wane and older teachers depart or lose interest. Many teachers reported on the ways in which people seemed to just come and go, meaning that friendships were established only in passing and were easily altered and re-established:

> Well you know, the English teacher population here comes and goes. I used to have a lot more British friends, but they kind of left and went back home. So nowadays, most of the people I hang out with are North American. (Eddie, Canada, Male, English teacher)

The churn of friends evident in Eddie's quote was a common articulation amongst English teachers who had been in South Korea for both long and short periods of time. Eddie was entering his second year of teaching and had been playing with the idea of learning more Korean language so that he could build relationships with people who were not necessarily going to leave, 'like they are not just going to up and leave, if anything, I am the one that is going to leave.' Margaret (UK, Female, English teacher) and Thomas (Canada, Male, English teacher) commented on a similar dynamic and the limitations of English teacher social networks:

> You can be really good friends with them, but you know one day they're gonna go back to their home country and, you know…You can call them close friends now, but maybe in a few months then you're not gonna see them ever again, but you're still gonna be in contact. [pause] That's why I want more Korean friends [laughs]. (Margaret)
>   A lot of people view Korea as a very temporary thing, it's like a working vacation. They're here for a year or two so if you make foreign friends, they're gonna leave anyhow, so most of my foreign friends are just people I meet at work. Right, and they're in and out, in and out, it's more of a, just a work friend kinda thing. (Thomas)

Like the general churn of teacher arrival and departure and the structuring of the migration regime around notions that teachers are on work–travel excursions, the dominant social practices amongst English teachers generate a sense of transience that influences even those who have remained long-term. It is to the question of generating permanence in a context of precarious–privilege that Section 6.4 turns.

6.4   Generating Permanence

The temporal patterns of migrant lives often hold individuals at a distance from stabilising relations with the places they inhabit. This is particularly the case for migrants on time-limited visas like workers, teachers and students where different forms of transience and turnover influence the ability to advance careers, build social relations and establish attachment to places; the temporalities of migration help constitute the position of individual migrants both in relation to resident populations and vis-à-vis other migrants. In the case of English teachers, their position as migrants in Seoul is generated in the particular expressions of desire that bring many to South Korea (to pay back debt, travel, earn money, to escape crises), the public temporal norms in their home countries that establish this form of migration as acceptable for a certain duration (following study as a work–travel experience but not as career), and in the governmental framings of English teachers in South Korea as necessarily temporary subjects. In each case these forces militate against longer-term temporal horizons within which teachers might understand themselves as having a place in Seoul, wherein they might stabilise social, economic and possibly even political relations in order to make a future life in the city. Unlike EPS workers, the constraints on teachers are not blunt durations established in legal restrictions but rather are constituted in the cross-cutting of desiring-migration, temporal norms and governmental framings. Allan (Canada, Male, English teacher), who had been in South Korea for over a decade at the time of the interview, put it this way:

> In the short term, yeah. You can live in this city. You can live in your own little world, work your job, make your cash and then when it's time to leave, you leave. But you're not staying. This isn't going to be your home but that's not a pattern for making this your home. It just becomes a place, a place to work and then eventually a place to leave.

The uncertainty expressed here by Allan emerges in relation to what can be described as a condition of 'permanent temporariness' (Bailey et al. 2002; Collins 2012) wherein individuals hold temporary social or legal status for extended periods of time. Amongst both English teachers and EPS workers this sense of time and the future was apparent. There were many English teachers who saw their work in Seoul as relatively short-term and intended to remain only for one or two years; similarly, many EPS workers saw their time in Seoul as clearly limited by the end of their work visas. As we have seen, however, in both cases there are individuals who remain beyond these shorter periods, crafting more permanent relations with the city and in work and social life. Chapter 4, for example, highlighted how some EPS workers become undocumented in order to remain long-term and develop alternative possibilities for themselves in the city. In the process such individuals place themselves at risk of deportation and potentially but not always more exploitative workplace conditions but they also assert an agentive will to exceed what is expected of them in the migration regime. By contrast, English teachers have no such strict time limits to subvert, they can remain indefinitely so long as they have employer sponsorship. Nonetheless by seeking to remain in Seoul English teachers are also transgressing expectations established in the migration regime and in wider social discourses and relations that their residence in South Korea will be necessarily temporary and tied to particular life stages.

In this final section I explore the narratives of teachers who remain long-term and focus on the practices of reconstituting subjectivity as part of efforts to generate greater levels of permanence in Seoul. While most teachers only remain for twelve months there is also a considerable contingent of teachers who have been resident in South Korea for many years. While general figures are not available, 12 of the 41 interviewees in this research had been in South Korea and/or Seoul for at least five years at the time of the interviews. Amongst these there were some participants who had been in South Korea for well over a decade. Except in those cases where individuals have taken up other forms of residence through marriage, these individuals have in effect become long-term non-citizens of South Korea, a population who are not undocumented in the way that some EPS workers can become but have no specific rights beyond the temporal duration of their contracts. As Landolt and Goldring (2015) argue, non-citizenship of this kind is a special relationship with the state, and I would argue in this context also with urban society, that relates not only to rights and rights-claiming but also to sociability and labour as constitutive of the everyday presence and practices of non-citizens. In this last section I focus exclusively on the narratives of these teachers who have become long-term non-citizen residents of Seoul and the particular ways in which partial and contingent expressions of permanence are generated in their lives. While there are multiple expressions and enactments of permanence in teacher lives I focus here on two relatively common examples across the interviewees, intimate relationships with Koreans and online cultures that emerge in a localised Anglophone blogosphere.

6.4.1   Intimate relationships

Amongst English teachers, intimate relationships with Korean men and women were relatively common, and much more common than they were amongst either EPS workers or international students. These relationships varied considerably, from very short-term flings to relationships that evolved into domestic co-habitation, marriage and in a few instances children, and of course separation and divorce sometimes following that as well. They speak to the complex forms of desiring that emerge in migration (Cheng 2011), not only desiring a place in the world but also closer relations with certain bodies and to the ways in which particular liaisons are deemed more or less acceptable in society.

Such relationships are also embroiled in highly gendered and racialised discourses about the appropriateness of relationships between Koreans and non-Koreans. The common trope is that English teacher–Korean relationships are bound by a heteronormative, male-teacher and female-Korean norm that is associated with undesirable elements of English teacher sociabilities. In public discourse the derision of these relationships has emerged at certain moments as a kind of moral panic about the negative influence of English teachers in particular but also western culture more broadly on young people. In 2005, for example, the prominent job posting and forum website 'English Spectrum' became a focus of considerable public debate after images were posted there of a costume party involving western men and Korean women where considerable emphasis was placed on the degrading of Korean women and the problematic lifestyles of English teachers (Wagner & Volkenburg 2011). Several female English teachers also commented on their dislike of these cultures and their effort to avoid the social spaces where they are reproduced:

> The thing is – like it doesn't feel … and with the crowd I hang out with. It's like a safe environment but in Itaewon the people are just crazy … they're just … You can see, at

the like cake house, whatever the bars. They're like … they're not normal, well they for one thing, like they will take the girls home. You go see. I don't like it that's all. So, I don't like it at all. (Chloe, South Africa, Female, English teacher)

While public imaginaries of intimate relationships with English teachers are dominated by this particular cross-cultural heterosexual norm, the narratives of teachers in this research revealed that relationships with Koreans were common for men, women, heterosexual and homosexual participants. In many instances, intimate relationships provided scope for teachers to feel more connected to South Korea or to specific communities in Seoul. The establishment of relationships gave participants additional insight into Korean culture and opportunities to meet a wider range of Koreans rather than only English teachers or other non-citizens. Gareth (Canada, Male, English teacher) explains:

I found with her, because of the kind of person she was, I was sort of much more plugged into what was going on in Korea, right, and it actually sometimes made relating to foreigners here difficult in a way, because they were sort of dumb-asses talking about what they didn't understand. … was hearing a lot of quite interesting, multiple viewpoints on that kind of stuff from her. And I don't know, there was a certain sense in which being with her sort of tampered my quick judgement of things, you know? … In a way I think sometimes being with a Korean, being involved closely with a Korean, even you know best friends, but especially in a romantic relationship, it gives you a much stronger stake.

Gareth speaks here of having a greater feeling of connection to place as a result of these relationships, a 'stronger stake' that generates a sense of becoming part of wider social relations. In addition, other participants, particularly the women who were in relationships with Korean men spoke about the legitimacy that could be established through their relationships – a claim to have an appropriate reason to being in South Korea rather than just an English teacher who had no other option. Mary (Canada, Female, English teacher) who was married for five years and had a child but then divorced from her husband captured some sense of this expression of permanence:

When our marriage was bad, it made my life pretty much a hell because I felt like I didn't have any reason to be here anymore on top of everything else but generally, I can say that having a Korean spouse gave me a lot of like, leverage I guess or a lot of legitimacy. And still does. I still say things like 'my husband' to people it just slips out… Having a Korean spouse, especially a Korean man, not a Korean wife but having a Korean man, as my husband is, it just makes things more legitimate, makes me seem like a more, 'oh I must really like Korea, I must not be as crazy foreigners are'.

Mary's explanation of the legitimacy generated in intimate relationships also speaks to their gendered character as expressions of permanence and connections into Korean society. Partly because of the gendered and racialised tropes that circulate around male-teacher and female-Korean relationships, legitimacy is not necessarily as accessible for male foreign teachers, unless they undertake to learn Korean and become embedded in society in other kinds of ways.

Relationships can also generate a bounding of feelings of permanence. Teachers noted that while being in a relationship helped to 'feel secure' (this was a common articulation) it also meant that in some instances the partner became the conduit for relationships with place, in the use of language, establishing friendships or making decisions. Francesca (Australia, Female, English teacher) and Margaret (UK, Female, English teacher), for example, both mentioned that partly through their relationships they stopped meeting new Koreans aside from their boyfriends' friends and had less imperative to learn and use Korean language on a daily basis because they would often go out socially with their boyfriends. In the case of two gay men in this research who were long-term residents of Seoul there was a sense that having long-term relationships had different kinds of effects, generating some security but also closing off access to the wider gay community where they had felt considerable belonging. They socialised once they were in a relationship but unlike heterosexual couples were not introduced to family networks because their partners had not come out to their families. For Martin (Australia, Male, English teacher) this 'actually made me less attached to the place' because it closed off involvement in community. For Jonathon (Australia, Male, English teacher) it involved a reconfiguration of subjectivity such that different relationships could be balanced and kept separate:

> I think I have integrated probably to the extent that I need to in the sense that I have a Korean partner, I speak the language and have my Korean life in the evenings with him. During the day, I have my more foreign life, but I am still interacting with my students. So I would say that the majority of the people I interact with during the day are Koreans rather than other foreign people. So I am integrated enough. I don't need to act like a Korean, that is not what it is about. It is understanding their view points, but being free of that and being somewhere in-between the foreign and the Korean world.

Intimate relationships as described here clearly draw teachers into new social formations and expressions of self. They create the possibility for thinking of greater levels of attachment to place in a context that is otherwise characterised by relative impermanence and precarity. As some of these examples also suggest, however, relationships are not simply a direct connection into Korean society but rather a reconfiguration of subjectivity and relationships that draws teachers into certain kinds of social formations that themselves are bounded and striated in terms of gender, race and sexuality. Intimate relationships generate the possibility of permanence but not an undifferentiated kind of inclusion.

6.4.2   Blogosphere

These intimate relationships involve generating connections into a more Korean-oriented life. There are also several 'community-like' activities that some longer-term English teachers engage in that have generated a sense of place, if not in the way that intimate relationships do. This includes the establishment of music, sports and church groups, the brief formation of groups like the Association of English Teachers in Korea, as well as the growth of a relatively vibrant blogosphere that I focus on here. Blogs have a relatively short history as a form of social media that has only grown substantially since the turn of the century. In the English-language community in South Korea too, the history is also relatively short with participants suggesting that the first blog Marmot's Hole was established in 2003, followed by a growth in numbers around 2006 and 2007. At the time of this research there were hundreds of English-language blogs written by non-citizens in South Korea. Initially the blogs were started by individuals who had been in South Korea for a considerable period of time, could speak Korean and hence used these platforms to document and translate news into English or to provide general information to foreigners in South Korea (English teachers and others).

There is an important overlap here with internet-based blogs and forums initiated and managed by migrant workers, which include nationality based websites like the Filipino EPS Workers Association and Vietforum, as well as initiatives like Migrant Worker TV (MWTV) that broadcasts alternative media for and about labour migration in South Korea. Lee's (2012) account of MWTV is particularly relevant because it speaks to the ways in which these online connections are informed by the socio-legal position of migrants in relation to society. Indeed, she notes that the emergence of MWTV and other migrant media in South Korea can be associated with the recognition of class connections across nationalities and the shared interest in supporting other workers and challenging negative portrayals that emerge in mainstream media. The blogs run by English teachers in this research serve some of this function in that they reflect a realisation of shared subjectivity amongst teachers in South Korea and specific sets of activities that allow for meaningful social relationships and addressing specific problems via the circulation of information. However, unlike MWTV they are, as we will see, relatively inward looking even by the accounts of those who run these blogs – they do not speak back to Korean media or public discourse, partly at least because many of the teachers involved do not have the linguistic ability to do this.

Half of the longer-term teachers in this research had their own blog and they were to varying degrees all involved in the nascent blogger 'community' in Seoul and in a more dispersed sense across South Korea. For some teacher-bloggers, blogging was an opportunity to filter potential social contacts in terms of potential shared interests and level of experience in South Korea. In many instances, blogging and the offline social lives that accompanied it involved avoiding the transient spaces of English teacher social lives:

> I don't go to foreigner bars, I meet too many weirdos there. So, the bloggers are kinda safe. And it leaves a bit of a soap opera there. How people intertwined, there are public feuds....I like them because all these people are just really interesting, they're very well-read, articulate, you know, they have ambitions ....They're not like over-ambitious but they're not 'I just want to spend my life at the foreigner bar, you know, do my school and get drunk every night'. (Jason, USA, Male, English teacher)

As Jason and several other interviewees put it, the blog space provides an alternative more stable world to the sociability that has become normalised amongst some English teachers. In part, blogging provides opportunities for individual teachers to develop other interests that are not solely related to their temporary stay or to passing commentary on cultural differences in South Korea. As Mary (Canada, Female, English teacher) put it, it also provided access to new social networks that were more concentrated amongst longer-term residents:

> I have a blog as well and that's how I got to know him [another blogger] and a couple of other people. That's probably the only social networking or socialising that I do or I've come to do in Korea now. I have had a lot of friends here but they leave. Not everybody's stayed for 10 years. Or I do have friends but they live in different cities. So, for the past year since I've been here, and even when I was in Daejeon, most of the people I socialise with I've come to know from websites.

The Anglophone blogosphere has emerged through a range of different channels and has come to take shape around particular personalities and networks. Some interviewees like Jason (USA, Male, English teacher) saw a mini celebrity culture emerging and a site wherein social networks could be reinforced and forms of differentiated inclusion emerged. Simon (New Zealand, Male, English teacher) explains:

> I think there is definitely a smaller clique, for want of a better word at the top, that is your Marmot's Hole, and your ROK-drop, and maybe there is a blogger in Gwangju, who I think revels in sort of being labelled as an angry dude out in the country.... Since being on this podcast since last year, I have sort of been allowed into this top, allowed to sit at the top table a little bit, just a little bit.... I'm just some Kiwi who writes a blog so his Mum knows he's still alive, and passes a few sarcastic comments on this podcast.

While the blogosphere then creates connections amongst longer-term teachers, several interviewees, including those who did not have blogs themselves, commented that it also risks insularity. While there are hundreds of English-language blogs amongst non-citizens, only a handful are well established and within that a smaller number have commentators who are connected socially beyond the English teacher or other non-Korean communities. In this context the blogosphere can be something of an 'echo chamber':

> It is an echo chamber and people in echo chambers have to think of it as a community because they'll go crazy if they don't. … I get very nervous when people start talking 'bout community, you know what I mean, especially with there's such a high rate of turnover, there's such a shallow degree of connection most of the time. I mean I feel much more a connection to, there are science fiction writer friends that I have in Australia or in Los Angeles which I feel a much deeper connection to because we're invested in the same things. … I don't know who said it, but someone said on a podcast recently 'Most of us here … many of the people who are here are running away from something' – I don't think that's a good thing to build a community upon.' (Gareth, Canada, Male, English teacher)

Online spaces like those generated in the Anglophone blogosphere have a particular kind of boundedness, what Gareth here refers to as an echo chamber. Situated in the imminently connected and extensive spatialities of the internet they articulate with wide networks but the particular social formation and its focus remains relatively narrowly focused on English teacher and other non-citizen lives in South Korea. It reflects in this sense a desire for connectivity and the consolidation of a particular migrant subjectivity around being an English teacher or more broadly a westerner in South Korea. The range of actors heavily involved is also relatively few, particularly in terms of long-term involvement. As a result, the kind of permanence generated here is generated around relatively constrained social networks of English teachers. The blogosphere has in this sense contributed to the formation of a kind of parallel long-term community of residents, most of whom have been or are English teachers. In many respects, it mirrors the formation of nationality and regionally specific associations amongst EPS workers and other migrants, although in this case with fewer active participants. In this respect, it is perhaps less effective at generating a sense of permanence because it relies on the ongoing presence of important actors or the arrival of new ones, rather than being part of establishing individual or collective relationships beyond the boundedness of English teacher or other migrant lives.

6.5   Conclusion

The figure of the English teacher is produced by a wide range of forces, not only independently as a category of the migration regime but also in relation to other categories such as migrant workers and international students. While immigration controls and visa categories are clearly important in the generation of this migrant subjectivity it is also clear that there are specific social, geographical and historical conditions that also help to produce the English teacher. Racialisation is particularly influential. The capacity to be an English teacher is associated with nations that are socially constructed as white and western and individual teachers' experience of employment and social relationships are shaped by their ability to align with this identity. There is a stark contrast then, that emerges most particularly with migrant workers as another prominent migrant figure in South Korea – they are presumed to be Southeast and South Asian and as a result experience quite different sorts of reception in the workplace and everyday life. Individuals who do not neatly fit these ascriptions of migranthood, whether teachers, workers or students, routinely reported either misrecognition, invisibility or aversion in their day-to-day lives. Read in this way, English teachers, or at least those who can align with expectations of desirable whiteness, are relatively privileged as migrant subjects in Seoul.

The discussion has however also pointed to other forces involved in assembling the subjectivities of the English teacher that has bearing for our understanding of the place of migrants in the city. Migration categories and racialised representations of migrants including English teachers are also articulated in relation to the individual and shared circumstances that generate migration and the conditions that shape employment and social relationships in the city. In the case of English teachers, their privileged status as white professionals is also attenuated by migration that often occurs in contexts of underemployment, debt and personal crises – migration to Seoul is rarely goal directed in this instance, it is rather about survival strategies and seize-the-moment opportunities. In daily life too, English teachers are clearly held apart from their colleagues and while they can be received positively by neighbours are also interpreted as short-term sojourners rather than potential long-term residents. This transience and churn clearly also influences how individuals position themselves in relation to the city, particularly those individuals who remain over many years who despite their experiences appear to be held in suspension between long-term residence and always imminent but never arriving departure.

In this respect, the accounts of English teachers in Seoul discussed here point to a curious alignment of privilege and precarity in migration as well as everyday lives in the city. The comparison with the lives of EPS workers is rather instructive in this case. Chapter 4 hinged on the recognition that the lives of migrant workers are characterised by marginalisation and the spatio-temporalities of the urban periphery: invisibility, regimentation in the work place and the importance of the mobile commons as a social support. Such a reading aligns with a significant body of research on labour migrants in Asia (Wong & Rigg 2010) as well as specifically in South Korea (Seo & Skelton 2017). English teachers do not occupy these same spaces of marginalisation. Nonetheless, the narratives presented here point to the ways in which the seeming privilege of English teacher migration can cover over questions of precarity that are significant features of their everyday lives in Seoul and potential futures in Seoul or elsewhere.

There is precarity generated in the ostensibly privileged migration of English teachers, particularly for longer-term teachers who express uncertainty about the future despite years or decades as non-citizen residents of Seoul. For those who remain long-term it is the establishment of relationships with Koreans and the formation of communities of English teachers that hold potential to become more permanent fixtures in the city. The outcomes can vary. As the examples of relationships and the blogosphere demonstrate the reconfiguration of subjectivity can involve feelings of connection and attachment to Seoul and South Korea, increased legitimacy and a greater capacity to manage life. But they can also take shape in feelings of compartmentalised life and the generation of social distance from others as communities emerge parallel to rather than integrated with other parts of urban society. Like those tactics of recognition discussed in Chapter 4 and the initiative of international students in Chapter 5, these undertakings reveal how migrants can rework subjectivities that are shaped by migration regimes, employment and institutional relations and social interactions. Desiring these alternative configurations of urban life exceeds what is expected in the political rationalities of contemporary migration, even if their actual effects vary considerably. As Chapter 7 will demonstrate such reconfigurations of subjectivity also influence the encounters that different migrants have, the impact of their social lives on the city and the coupling and decoupling of migrant and urban futures.

Endnotes

1   Marmot's Hole closed at the end of December 2015 as this chapter was being written. The author Robert J. Koehler had decided that he no longer wishes to take part in the debates that dominate the blog space and has started a photography blog instead. He continues to live in Seoul.
2   Sunwoo, C. (2013). Korea by blog – Creating an expat blogosphere. *Korea Joongang Daily*. Available at: http://koreajoongangdaily.joins.com/news/article/article.aspx?aid=2968810 [accessed 18 Dec 2017].

Chapter Seven
Multicultural Presence and Fractured Futures

When I think of the future of Seoul as that of a 'world-class city', I conjure up the image of a foreigner enjoying a beautiful cityscape from his or her window, or running alongside a clean Cheonggyecheon. Our ultimate goal is to make a cleaner and more livable city so that someday we hear foreign workers say, 'Seoul is the place I want to work!' Former mayor of Seoul (2002–2006), Lee Myung-bak (Seoul Metropolitan Government 2002)

Alongside the growing significance of migration in South Korea, there has been heightened policy emphasis on reconstructing Seoul as an urban space of flows, connected to and embedded within regional and global circuits of economic and cultural possibility (Watson 2012). The connections to migratory flows are captured evocatively here in former mayor (2002–2006) and president of South Korea (2007–2011) Lee Myung-bak's imaginary of Seoul as a desirable place to live and work not only for Koreans but for 'foreigners' and 'foreign workers'. Aligning his administration's emphasis on the regeneration of key urban zones like the restoration of the Cheonggyecheon Stream with the effort to attract flows of talented migrants, this assertion can be read as expressing desire for global city status. Lee's successor as mayor, Oh Se-hoon (2006–2011), articulated similar ideas when he claimed that 'Seoul is transforming itself into a place where diverse peoples and cultures come together…a true capital for all the peoples of the world' (Seoul Metropolitan Government 2008). Accompanied by substantial investments in place marketing, a network of Global Centers 'to enable the foreign citizens of Seoul to lead a hassle-free life', foreign-owned business

Global Asian City: Migration, Desire and the Politics of Encounter in 21st Century Seoul, First Edition. Francis L. Collins.
© 2018 John Wiley & Sons Ltd. Published 2018 by John Wiley & Sons Ltd.

incubation, global business zones, international schools and housing, these discourses speak to attempts to harness particular kinds of diversity for urban transformation.

These discourses, however, are also marked by silences that reveal the segmented character of desiring global cityness through migration and diversity. Clearly, given his reference to the gentrified urban-nature spectacle that is Cheonggyecheon Stream (Križnik 2011), Lee is not thinking here of a future Seoul peopled by foreign guest workers arriving through the EPS or even possibly English teachers from western countries, but rather of a more desirable 'talent' migrant. Oh Se-hoon similarly saw the achievement of a multicultural city in the ability to 'attract global talent wishing to take advantage of the opportunities the city has to offer'. International students like those attending KU and SNU hold potential to become these talented migrants (Shin & Choi 2015), but they too are not immediately the concern of or themselves concerned with business zones, international schools and specialised high-cost foreign housing. Rather, each of these significant streams of migration are to varying degrees marginalised from the vision of becoming a *Global Asian City*. The multicultural presence that is proposed to constitute the future global city is not all-inclusive but rather focuses on the discrepant desirability of migrants and their capacity to participate in elite urban spaces.

These urban discourses of aspiration, diversity and global status overlap in curious ways with a growing national discourse of *damunhwa juui* or multiculturalism that has become a pervading feature of politics, culture and society in twenty-first-century South Korea (Lie 2014). Responding to the growing number of non-Korean citizens residing in the nation, discourses of multiculturalism also speak to notions of transformation and evolution in nationhood. Former president Roh Moo-hyun (2002–2006), for example, famously declared in 2006 that '[i]t is irreversible for Korea to move towards a multiracial and a multicultural society. We must try to integrate migrants through multicultural policies'. He would follow this announcement with the establishment of a Committee for Foreigner Policy and wide-ranging investments in research on and responses to multicultural challenges that have continued under the more conservative administrations that have followed (Lim 2014). Like the urban discourse of diversity as talent, these national discourses are also characterised by a segmentation that is revealed in their silences. As numerous scholars have maintained (Kim, A.E. 2009; Kim, J. 2011; Lim 2010), multiculturalism in South Korea has been almost exclusively concerned with the assimilation of female marriage migrants to the social and cultural norms of Korean families and society. The 'irreversible' assembly of a 'multiracial' and 'multicultural' nation, then, has been premised on the ability of women married to Korean men to fulfil their expected role 'as mothers of the next generation of Koreans' (Kim, M. 2013: 456).

EPS workers, English teachers and international students appear to have no formal place in these impulses towards national multiculturalism. Rather, their inclusion has occurred through migration regimes that seek to differentially manage their presence and future prospects. As I detailed in Chapter 3, these regimes establish the roles expected of different migrant groups – as time constrained workers in the urban periphery, short-term educators on work and travel experiences and contributors to the diversity of university campuses. As Han (2015: 3–4) puts it, 'Korean multiculturalism is preoccupied with providing "new Koreans" with their needs and effectively controlling "the others" as marginalized members of Korean society'. Clearly, the level of marginalisation varies in material terms but Han's concern here is with marginalisation from the prospect of becoming Korean, or aligning oneself with the Korean nation socially and culturally. In that regard, present policies of multiculturalism and future visions of a multicultural society only view marriage migrants and their children as potential subjects of the nation. For others, there is a complex and multilayered experience of 'differential inclusion' (Mezzadra & Neilson 2012) where presence and prospects are guided by the production of varied migrant subjectivities. Workers, teachers and students are not excluded from the present territorial configurations of the nation or the city for they are very literally present and active subjects of those arenas. However, their absence from discourses and policies of national multiculturalism and urban diversity reveal how 'inclusion in a sphere or realm can be subject to varying degrees of subordination, rule, discrimination and segmentation' (Mezzadra & Neilson 2012: 67). Their exclusion is from imaginings of the future of the nation; they are expected to conform to the contours of their status as workers, teachers and students and to their consequent role and tenure within the nation and city. Put otherwise, the multicultural presence in contemporary Seoul articulates with fractured futures.

This chapter takes this characterisation of urban diversity in Seoul as a starting point for examining the way migrants conform to and exceed these expectations both in present practices in the city and the way in which their lives are differentially articulated in relation to urban and national futures. I focus in particular on the reality of multicultural presence in Seoul and South Korea and the manner in which this is articulated with the varied future projections of migrants themselves. In this regard, I shift focus from the previous chapters that have emphasised the present conditions of workers, teachers and students to draw attention to the processes of becoming that emerge in varying encounters with life in Seoul. To do this I draw the narratives of all three migrant groups together again, reasserting the importance of making 'concurrent those views and experiences that are ideologically and culturally closed to each other, and that attempt to distance or suppress other views and experiences' (Said 1993: 37). Drawing these accounts together makes it possible to observe both segmentation and alignment in the lives of different migrants and the manner in which their presence manifests in urban spaces. I begin through a discussion of the emergence of a more diverse urban politics of multiculturalism that both intersects with but also exceeds and at times subverts the normative goals ascribed by urban and national governments.

This includes the ways in which the practices of different migrants align with the consolidation of identity or subjectivity around the categories of migrant worker, English teacher and international student, or specific ethnicities and nationalities, as well as the way that some migrants push away from these assemblages and seek to build alternative relations and spaces in the city. Following these traces of multicultural presences, I turn to address the transformation of subjectivity in three ways that link present practices and future possibilities: the establishment of foreignness in migrant lives, strategies for aligning with Korean personhood, and the coupling and decoupling of migration and urban life in Seoul. The chapter concludes by returning to the politics of multiculturalism in Seoul and its implications for the crafting of a future *Global Asian City*.

7.1 Another Urban Politics of Multiculturalism

The urban aspirations expressed by Lee Myung-bak and Oh Se-hoon are examples of the dominant urban politics of multiculturalism. This is a politics that centres on the promotion of particular kinds of ethnic and cultural diversity, particularly that associated with desirable forms of migration (Han 2015). As Mitchell (1993) argued some time ago, these kinds of urban politics of multiculturalism are intimately linked to neo-liberal forms of globalisation wherein policy and discourse take hegemonic control of notions of race and nation. It involves a reterritorialisation or stabilisation of difference in order to enhance the symbolic and material desirability of cities. As such these politics are organised not so much around open spaces of encounter between people of different backgrounds, a deterritorialising impulse, but rather in segmented spaces and activities of formally recognised and regulated diversity such as festivals and marketplaces. In the context of Seoul and other major global cities in East Asia, these political rationalities also often align with national modes of managing migration that are directed towards attracting and retaining a mobile elite and maintaining the marginality of migrant populations (Ong 2007; Yeoh 2006).

The efforts of urban and national governments to coordinate the presence of diverse populations and the outcomes of migration rests on a presumption that differences exist prior to migration and can then be managed in the organisation of urban life (Darling & Wilson 2016). In addition to contributing to the kinds of discrepant experiences documented in preceding chapters, this political rationality of diversity also ignores the mutually transformative articulation of migration and urbanisation. As Isin (2007: 223, emphasis in original) explains, 'we need to recognize that the city is not a container where already formed differences (e.g. slave, craftsman, merchant, woman, warrior, bourgeois, queer) arrive in the city and encounter each other. Such differences are generated *in* and *through* the city'. To Isin's (2007) list we could easily add migrant worker, English teacher and international student in the case of Seoul and recognise that the differences ascribed to these groups are generated not prior to migration but in their articulation with urban life and in ways that intersect with classed, gendered and racialised renderings of different individuals. In what follows I seek to address two dimensions of the ways in which differences are generated in encounters in the city and in particular in the divergent social practices of migrants. I focus first on forms of social practice that involve a consolidation of particular identities around shared interests and practices before exploring instances of aversion from or avoidance of those consolidated identities and the formation of alternative social arrangements or impulses. Together, these examples reveal the limitations of thinking through the urban impacts of migration in terms of a politics of segmented and pre-existing difference in need of management and the importance of paying attention to the generation or assembling of different possibilities for migrant life in the city.

7.1.1 'It's the same every week like a circle'

Alongside formalised governmental imaginings of multiculturalism and urban diversity exist a whole raft of different practices that are *already* undertaken by migrants that reconfigure the character of urban spaces and create possibilities for particular kinds of encounters in the city. Workers, teachers and students described engaging in a variety of social activities beyond work and campus spaces, attending religious services, playing sports, dancing, photography, exploring the city, going to bars and nightclubs, communicating with home and hanging out with friends. The routine character of such activities is both prosaic while also highlighting their significance in terms of providing opportunities to be more than just migrant workers, English teachers or international students. This was revealed particularly starkly in the accounts of migrant workers who, due their dispersed location in the urban periphery and limited time off work, had to invest significant effort to go out rather than rest. Minh (Vietnam, Male, EPS worker) explains:

> I have two hobbies football and photography.... Most of the free time, I go to shopping malls, photography shop centre in Namdaemun or Yongsan. Or I visit old palaces in flower seasons or in the autumn like now.... It's the same every week like a circle. In the morning I get up late because of tiring working days, and then I will go somewhere. At first, if I decide to play football I will go to see some friends in the team from noon to 5 pm. When the match ends, the team will go for a drink, drink beer and eat chicken while talking about work. Sunday ends like that. It takes me one and a half hour from home to Seoul.

Activities like photography and football are subtle but nonetheless important appropriations of public space that involve individuals directing their lives in relation to their own interests (Elsheshtawy 2008). They can also emerge through forms of sociability that cultivate familiarity with others and a sense of belonging that can often be undermined in contexts of transience and churn where people are coming and going. Gareth (Canada, Male, English teacher), for example, compared his experience between years when there was a lot of turnover at work and when things became more settled around socialising patterns:

> So it was a little bit isolating and also at the time that I arrived there were a number of dysfunctional foreign teachers at this school, so we didn't really have a strong community within the foreign teachers on campus. This year we have a really good one actually, you know a lot of socialising and we'll go events and things like that, which has helped the new people to sort of adjust if they came in, especially from outside of Korea, there were these here for them to plug into that, you know a little social circle.

As Gareth's account and the routine activities described by Minh suggest, socialising often revolves around the consolidation of social networks of familiarity or similarity, the creation of a 'strong community'. Most participants in this research described spending time with other migrant workers, English teachers and international students, an arrangement that emerges through co-presence and shared circumstances, or indeed the patterns of arrival and adaptation discussed in previous chapters. Priya (Sri Lanka, Female, SNU Student), for example, described how her social networks and practices emerged through early experiences at university:

> We had 10 students, scholarship students, and there were three students from El Salvador, and like I said, two from Sri Lanka, two from Sweden, Bhutan and Angola. Those were the 10. ... For us, because we had our own [Korean language] class, they started from the beginning with that. And everything. So, it was fun, because it was just international students, and we were all new, and we, we would like go out to eat, and like – explore, and stuff like that. So, it was really fun. Exciting.

Rather than prefigured social networks based on migrant status or nationality, what matters more here is the specific sites within which migration and urban incorporation occurs. Priya's account points to the generation of shared experiences, alignments of different individuals in the classroom and subsequently in the ways in which individuals inhabit urban space more generally, eating, exploring, going out. While there is much socialising across nationalities it was clear that alignments of different bodies can also consolidate around ethnic, national or linguistic configurations. Bayani (Philippines, Male, EPS worker), for example, talked about instances of socialising with Korean colleagues but the seeming difficulty in aligning practices and affinities:

> I know quite a lot of foreigners, like the Koreans I work with, so we become ... I can't really call them close friends, but they're acquaintances. I go out with them from

time to time....We drink, since they're fond of drinking, as you know. I hang out with them, I go see new places with them from time to time. But since that seldom happens, and since of course there's the FEWA association [Filipino EPS Workers Association], of course you end up choosing to go out with fellow Filipinos instead of Koreans, right? [chuckles]

This coherence of social practice around ethnic and linguistic differences should not be surprising given that many participants in this research had limited grasp of Korean language and were subject to forms of differential inclusion that meant their daily contact was mostly with others in similar situations and sometimes only co-nationals. Desiring enters into the making of social relations in the city here, as we can observe the pull towards familiar forms of embodiment, language, behaviours and activities, or amongst people who share similar routines and circumstances such as the absence of family.

The effects of these socialising practices centred on shared migrant status and co-ethnic or co-national affiliation manifest in appropriations and transformations of urban spaces. In Seoul, there are several urban spaces where the presence of migrants, particularly migrant workers and English teachers, has become apparent in weekly routines to shop, socialise, call home, bank, attend classes, send goods, haggle and avoid the social and cultural norms that reinforce their status as migrants and foreigners (Yun 2011). In the case of migrant workers there are two well-known examples: a retail strip in Wonggok-dong (Ansan) referred to as 'the borderless village', a destination for a wide variety of nationalities including Indonesians, Vietnamese and Chinese (Yun 2011); and the cathedral and small Filipino street market in Hyehwa-dong (Yea 2015). As Mindo (Philippines, Male, EPS worker) suggests, places like Hyehwa become the normative destination for migrants not only because of the potential for familiarity and social contact but also because they are known destinations:

> Where do I go out usually? Just here [Hyehwa]. The truth is, when I was here for my first year I was always just here. Every Sunday I just stayed here. It became my home. Because I didn't know what time our bus would pass. So even now, I can I usually stay out till the evening, but I just [stay] here....During Sundays, I don't think of [other places] When you go here, your time gets used up.

Places like Hyehwa and Wonggok-dong have a vibrant marketplace-like atmosphere for several hours every week as migrant workers gather to meet friends and go about activities that they otherwise have little time for. As scholars exploring migration in other Asian cities have suggested, these urban spaces can be understood as tactics for 'defying disappearance' (Law 2002; see also Yeoh & Huang 1998). They provide opportunities for becoming ordinary residents but in the process they can also involve a movement from the periphery to more prominent urban spaces, a reconfiguration of migrant life from those places that are made invisible to spaces where aspirations for world city status and the politics of aesthetics are most explicit. Accordingly, they have also been subject to governmental projects to craft more ordered iterations of multiculturalism. The grassroots moniker 'borderless village' has, for example, been rebranded in official terms as the 'Multicultural Street' with murals and street signs that celebrate the presence of different populations (see Figure 7.1). While certainly not an instance of elite urban space making, there still appears here a desire to mark out difference in the urban environment in ways that it can be known and perhaps then also contained. More striking was the effort in 2010 by city officials to close down the Hyehwa-dong market as part of a broader campaign to remove vendors from Seoul's streets, particularly in the important business and retail districts of the city. The district government attempted to move the activities to a formalised 'Multicultural Street' in another part of the central city, Nakwon-dong, a move that aligns with efforts to rebrand Seoul and redevelop its older inner areas. The plan to remove the market was actively resisted by the church and migrants and the market subsequently received explicit support from the then new mayor of Seoul Park Won-soon who has a notable history in human rights activism.

There are considerable similarities that emerge in the ways in which English teacher socialising practices take shape in relation to specific urban spaces that

Figure 7.1   The official 'multicultural street' of Wonggok-dong.

then become associated with their presence in ways that affect both teachers themselves and form part of the representation of areas. With time on their hands and for some their first taste of regular income, the social lives of teachers can be lively, particularly for those in their first few years of teaching. Vibrant nightlife areas like Itaewon, Hong-dae and Sinchon are regularly identified as the places to access familiar goods and services but also to meet other foreigners and to step outside of the daily focus on work. Cindy (USA, Female, English teacher) described her feelings around the well-known foreign district of Itaewon:

> I don't feel like I am out of place because there are enough foreigners that make the area come alive.... I go there partially because that is where many of the international restaurants are, because it's where a lot of the culture you are first introduced to when you come to Korea, that is where that really is. It's a great place to hang out with foreigners because it is known, that was a reputation I found as soon as I got here. Itaewon was the place to meet foreigners.

Like the neighbourhoods of Ansan or Hyehwa described above, places like Itaewon where teachers spend time also become associated with foreignness, more often than not whiteness, especially on weekends. They are spaces reconstituted by particular forms of language, drinking, eating, music, dancing, flirting and friendship. They are marked here by Cindy in contrast to other parts of Seoul as a place of culture, here clearly meaning culture that is not restricted to Korean forms but rather is associated with 'foreigners' and being 'international'. Itaewon has a long history as a 'foreigner district', located adjacent to key military installations of the Japanese and American forces for over a century. It has over time acquired a representation as a problematic space of foreignness associated with military presence but also moral concerns about relationships between foreign men and Korean women as well as providing a space for Koreans seeking escape from social and cultural conformity (Kim, E. 2004). More recently, Itaewon has been subject to processes of gentrification as the forms of foreignness expressed here have become associated with desirable modes of diversity. This has manifested in investment in new streetscapes, branding initiatives and a changing retail and hospitality landscape that has seen the growth in more high-end restaurants alongside a growing diversity of food provision beyond western or generically international cuisine (Song 2013). While not necessarily displacing English teachers, these changes can also be seen as part of efforts to order diversity in the city in ways that clearly respond to and seek to contain its more emergent and excessive forms.

There were no equivalent appropriations of space that emerged in accounts of international students and nor are there any areas in the city outside of university campuses that have representational associations with students as a population of migrants. Of course, international students engage in multifarious social activities that are significant in creating spaces of familiarity and attachment, including playing sports, learning language, attending religious services, eating, drinking and socialising. Two factors seemed, however, to shape the impact of their presence in the city. First, many students reported having little time and money to socialise – those who were self-funded and not working in particular found the cost of living in Seoul required them to limit their social activities more so than both workers and teachers.

Second, however, as Priya's example earlier suggested many of the socialising networks of international students were across nationalities that developed in the arrival and enrolment patterns. International students, like workers and teachers, are in similar situations to each other and experience some distance from their Korean peers at university because of language and differences in social networks and practices. The multicultural dimension to social interaction was notable however, even amongst students such as those from China for whom friendship networks included significant numbers of co-nationals. In other words, the particular configurations of migrant lives matter for the ways in which they come to inhabit and potentially appropriate urban spaces, for their will and capacity to rework or reterritorialise spaces according to their own desires. In the case of students, while they certainly engage in socialising practices and do so in specific spaces in the city their presence is not as visibly associated with particular ethnic and national associations in the way that places like Itaewon are connected to whiteness or the association of Hyehwa-dong with the presence and practices of people from the Philippines.

7.1.2   'I don't want to be seen as one of those people'

Familiarity and stability are not always desirable arrangements in the way the accounts above suggest. Indeed, while most workers, students and teachers pointed to the pleasures of common spaces and practices of socialising there were also participants who described places like Itaewon or Ansan in more negative terms, seeing them as indicative of the problems of migrant life and as contributing to distance between people. Not all teachers, for example, appreciated the Anglo-centric spaces and practices of Itaewon and foreigner bars, criticising them as reminiscent of 'crappy college time back home, drunk people being rowdy' (Emmanuel, Canada, Male, English teacher) or characterised by violence, sexism or racism. Such spaces, as Heather (New Zealand, Female, English teacher) suggests, were also perceived to be something that disrupted other more meaningful relationships in Seoul:

> Around Itaewon yeah, because if you just go to any old club, drink beer, there is no particular thing that would make you connect to the club. A lot of those clubs are full of foreigners drinking beer and they come in for the weekend, are always drunk. I don't want to be seen as one of those people, I want more of a connection with my community than that.

Heather's comment points to the ways in which socialising practices contribute to the framing of different migrants, in this case English teachers, as problematic and her own desire not to be limited by those representations. Tuan (Vietnam, Male, EPS worker) made similar points about Ansan and the problems he observed in social practices there:

> Conflicts between foreigners can occur when they live here for a long time. Most of the conflicts involve fighting. Where I live doesn't experience as many social problems as in Ansan. There are not as many foreigners who live here as in Ansan. There are fewer factories here than in Ansan so foreigner fighting or negative things like gambling or lottery still occur but not as much as in Ansan.

For both Heather and Tuan there is a desire to disassociate from the problems associated with forms of social life consolidated around some migrant identities, to have different kinds of relationships to place. This is stated most directly by Heather when she argues that she doesn't 'want to be seen as one of those people' and that she 'want[s] more a connection with [her] community than that'; but it was also clear for Tuan that he wishes to constitute himself in ways that are not limited to only reproducing social norms that could be interpreted negatively by local populations.

At times, this aversion to the alignment of bodies, spaces and practices in the city can take on significant gendered dimensions. Each of the female migrant workers in this study described socialising within much smaller primarily female networks that often crossed over nationalities and sometimes included Korean women as well. Sabrina (Philippines, Female, EPS worker) talked about bonding with 'sisters' who were Filipino and Filipino-American, as well as a Korean woman who manages workers at her factory, 'She'll treat me for food anywhere. Places I've never been to before'. Dewi (Indonesia, Female, EPS worker) and Thuy (Vietnam, Female, EPS worker) provided similar accounts, indicating social networks with Vietnamese, Thai, Chinese and Indonesian women, including both workers and women who had arrived as marriage migrants. Seo and Skelton (2017) observe a similar tendency amongst Nepalese EPS workers in their research, between public and visible spaces inhabited and appropriated by men and more intimate social relations between women, which as these examples suggest can also cross nationalities. While more varied amongst English teachers because of the greater gender balance, some female teachers also remarked on the gendering of English teacher social practices, particularly in terms of the negative connotations of male English teacher–female Korean relationships that they observed:

> To that extent I would agree for example I found that most of the male American foreigners who were here have very conservative ideas about gender and of course I…you know I'm not really like an activist in that sense but I do chaff at the idea that it's good to marry Korean women, 'cause she'll be so submissive to you and all that. (Charlotte, USA, Female, English teacher)

It is important to note, then, that even within seemingly stable spaces consolidated around particular modes of foreignness there are a diversity of practices and micro spaces wherein other less dominant expressions of embodiment and relationships emerge. Itaewon, for example, also has an established set of LGBTQ connections that have been made possible because of its broader atmosphere of subverting social and cultural conformity in South Korea (Kim, E. 2004; Kim & Hong 2007). This mattered for individuals like English teachers, who because of their role in education and often with children felt that their jobs could be at risk if they were identified as being outside heteronormativity. Lauren (New Zealand, Female, English teacher; emphasis added) described the role of spaces within Itaewon in maintaining intimacy and relationships:

> Because we're gay, I think, *with certain Westerners or fellow foreigners here*, we can be open about it, so we're more relaxed? And also, the conversations we have are more free-flowing – at the Seoul pub people aren't going to remember and we hid the names of our schools, so it's safe, we can be like 'Oh you know, we're together'; no one cares. So in Itaewon we can also hold hands, and you know we're not pretending, and so I think in a big way, that kinda means that we like to limit how much time we spend with Korean people, because we're closeted with Koreans.

There is a complex configuration of space and encounter described here. On the one hand, it is the nonconformist and foreign associated character of Itaewon, and its distance from everyday life in Seoul's suburbs, that makes this space 'safe' for expressions of homosexual intimacy. Yet, echoing Charlotte's comments about the prevalence of conservative ideas about gender, even within this space Lauren maintained that is only with 'certain Westerners' that she could be open about her sexuality and that there was still a need to hide the names of schools. Small spaces within a place like Itaewon provide possibility for more open relationships, but even then, there are limits and constraints around what can be expressed.

The configuration of migrant subjectivity in relation to specific urban spaces is not stable or singular then, even in relation to generalised migrant categories like migrant workers, English teachers and international students. Rather, it takes shape in relation to individual desires, sense of place and the experience of different embodiments in the city. Unsurprisingly, there are also alternative configurations of social life in the city that were described by some participants – including participating in activities that actively cut across ethnic or national differences or distinctions between different migrants and Koreans. One such example of establishing different kinds of spaces and activities that reached out beyond the strictures of specific ethnic and migrant formations was the performance of music. A small number of participants talked about their experience of being in bands and playing in Korean venues, often through relationships with the owners of bars or cafes. Nadia (UK, Female, English teacher) and Wijaya (Indonesia, Male, EPS worker) provided two examples:

In Anyang there's a Westerners' bar, and it's pretty much the only bar that there will always be Westerners in, from all around Anyang. So, just when I came to Anyang, I went there and I met loads of people. And the owner, she's Korean but she's really kind of really 'out there'.... One day she just said 'I want to start a band, but I don't know how to play anything', and I was like 'Well, I'd like to play but I can't, I don't know how to play anything either'. So I just said 'I'd love to learn the drums', so she said 'Right, you're the drummer', she said 'I want to learn guitar', so I said 'Okay', so ... She just asked some other girls and so the four of us just learned how to play and you know it's just fun. (Nadia)

You know, we were stressful and tired. We wanted entertainment after we worked.... At Idola café. It was owned by Mr. Kim. He married an Indonesian woman too. ... He was the owner of the café. There was also another café, Bali Indah, it was owned by Mr. Seo. ... We just used the guitar. It is natural. We could play guitar without the notes. Well, it's '*dangdut*' [traditional Indonesian folk music]. We just need to listen. We could immediately play it. What is difficult about '*dangdut*'. (Wijaya)

Both of these examples show how individual teachers and workers become embedded in alternative spaces through musical performance and connections with Korean owners of bars and cafes. They point to the kinds of affective resonance that different kinds of performance can have in migrant lives (see de Dios 2016), shared instances of 'fun' and 'entertainment' that can act as conduits to alternative relationships with local populations and developing attachment to place. In the process, they influence the micro spaces of the city in different sorts of ways from the consolidated identities of English teachers and migrant workers described earlier. Rather than appropriate space around fixed identities, music here offers opportunities for these individuals to work with others who are not migrants in the performance of music and the joy that comes with that and also to generate alternative encounters with audiences who are not limited to people in similar circumstances.

Another example of more open encounters and the alternative politics of multiculturalism that can emerge in migrant lives related to undertaking volunteer, outreach and activist work. There were multiple examples of this, including Nonoy (Philippines, Male, EPS worker), who was involved in building a migrant coalition connecting workers and migrant wives in Cheonan where he originally lived and worked; Monty (Philippines, Transgender, EPS worker) was heavily involved in the Migrant Trade Union as documented in Chapter 4; and Margaret (UK, Female, English teacher) who started a programme to bring English teachers to an orphanage in order to provide English lessons and activities for children. Another example came from Dalton (Uzbekistan, Male, SNU Student), who had been the founding president of the SNU International Students Association (SISA):

We at SISA made a programme for, I mean, to help multi-cultural families like usually women from other countries ... and married to Korean men. Living in this area, to help them adapt to Korean culture and learn some language. And, to help them, you know, solve their … daily issues with local people. And [the international] office is quite active in helping us with gaining some information. And also, they have that monitoring programme to make things better. You know, they gather some Korean people from the area and international students who are foreigners, to share some good information or ideas.

The initiative described here by Dalton relies on a capacity and will or desire to reach out and build relationships with others that are not limited to the confines of international student, English teacher and migrant worker lives. Notably, however, these practices do not occur outside of the framing of international students as migrants. Indeed, students' recognition as knowledgeable and to provide 'good information' to marriage migrants reproduces the position of the international student as a more privileged and able migrant that aligns with their wider framing in migration regimes. So too, the opportunity to teach English to orphans works through the recognised and desired capacities of English teachers as Anglophone westerners. Nonetheless, these are not simply the politics of pre-existing, segmented and cumulative multiculturalism that is imagined in accounts of urban diversity in Seoul and national multiculturalism in South Korea. Rather, this is a sense of diversity that functions through the arrangement or assembly of different sorts of encounters, where forms of desiring different kinds of relationships to place or community and opportunities to become more than English teacher, international student and migrant worker are fundamentally important. In small but nonetheless significant ways, these practices constitute a different set of social relations in urban life that exceed what is expected in migration regimes and the urban politics of multiculturalism. Because they emerge through encounter, however, they also point towards the varying transformations in subjectivity and future orientation that are possible through life in the city, a point that I turn to in the second half of this chapter.

7.2   Migration and Becoming

The insights generated above demonstrate that the urban politics of migration and migrant lives 'cannot be reduced to the desired outcomes of specific modes of intervention' (Wilson & Darling 2016: 11). Migrants are not simply additive to the city in the way envisaged by political discourse of urban leaders like Lee Myung-bak and Oh Se-hoon but rather become discrepantly migrant through their encounters in the city and the different ways in which they inhabit or appropriate spaces. Migration leads to the transformation of urban spaces whether across neighbourhoods as individuals take up residence and work, negotiate connections with local populations and adapt those spaces or in more micro level appropriations that are less visible but can involve alternative possibilities for social relations and meaning in the city. There is, as the previous section has made clear, unpredictability to the ways in which encounter and difference manifest in urban life because of the diverse desires of migrants and other urban inhabitants and how they intersect with possibilities available in different times and spaces of the city – think Hyehwa, Itaewon, churches, orphanages, football; drinking, flirting, fighting, gambling but also teaching, photography and playing music; all of which emerge in relation not only to consolidated migrant identities but also divergent responses to the possibilities for urban life in Seoul by migrants themselves. In accounting for the emergent and less predictable dimensions of migration's urban implications we must also recognise that encounters in urban life are also transformative for migrants themselves. Encounters in the city present opportunities for learning, self-reflection, resistance and subversion, and new identity formations that have significance for both present and future possibilities in individual lives.

In order to grasp the transformative potential of migration, the future possibilities that it generates, it is necessary to explore the ways in which processes of becoming reach across the formal categories of migration status – worker, teacher, student. This involves returning more explicitly to Said's (1993) call for contrapuntal forms of analysis that explore seemingly dissimilar objects of analysis alongside and in relation to each other. The identities of migrant worker, English teacher and international student are obviously products of the migration regime rather than something that reflects any phenomenological basis of the migrants who people these categories (Rodriguez & Schwenken 2013). They are invariably unfamiliar subjectivities to the very people they apply to. As such, while these identities certainly have a role in drawing the lines of who can move, where they can go and what they can do there, they cannot determine the results of migration, either in terms of values ascribed or the way that people take up different identities. To address the transformative practice of migration means paying attention to the manner in which migrants are involved in the making of the new that works through while also surpassing the regimes that generate and modulate migration.

Because it is not possible to trace all the varying and overlapping lines of flight that emerge in the narratives of 120 migrants, I focus here on a small number of shared expressions that reveal overlaps as well as differences within these groups. I begin by outlining the sense amongst many migrants of becoming forever foreigner, of never being able to align oneself with social formations in Seoul and South Korea. This sense of distance and being distanced points to the ways in which encounters in everyday life can reinforce the ethnic logics that inform contemporary multiculturalism and urban diversity. Following this, however, I focus on the less common ways in which some people are able to configure their lives and orientations in relation to what appear on the surface to be rigidly policed and ethnically-scripted notions of Korean personhood. In contrast to a framing as necessarily foreign, the examples of individuals building relationships across difference reveal the potential for cosmopolitan openness in migration as well as the implications of inclusion in migrant lives. Last, I depart from the emphasis on socio-cultural encounters to consider the future trajectories of different migrants, the coupling and decoupling of their lives with the making of Seoul as a *Global Asian City*. Identifying these expressions of becoming necessarily involve artificial processes of categorisation on my part and they are certainly far from being mutually exclusive. Nonetheless focusing on how these particular becoming processes emerge demonstrates how different subjects come to align and disconnect themselves from people and place through the encounters and experiences that they have in the city. As the discussion will also show, these future possibilities result not from singular rational and individual choices but rather like biographies of coming to South Korea discussed in Chapter 3, reflect desiring interactions with migration regimes, encounters in place, imaginations of places beyond the present and the affective pull of family and obligation.

7.2.1   Forever foreigner

> Like I know that forever here, I'll just be looked as a foreigner. Like it doesn't matter how well the culture – like, how well I know the culture, the customs, anything. Like, I'll always just be the foreigner. Like if someone sees me on the streets, they'll be like, oh, he probably just got here today, or something. (Arturo, Mexico, Male, Student SNU)

Becoming self-conscious of one's foreignness and status as a foreigner was a widespread articulation by participants. While some participants felt that they could 'pass' as Korean, and others suggested mechanisms for actively aligning oneself with Koreanness, the common expression was that migration involved becoming a foreigner. This is not surprising given the differences encountered generally in migration but also relates specifically the pervasiveness of myths of ethnic homogeneity since the twentieth century on the Korean peninsula (Lie 2014). The notion of Korean personhood rests exclusively on an interpellation of Korean ethnicity and lineage that demarcates and distances that which is perceived to be foreign. It also aligns with a view of multiculturalism as the incorporation of only select groups like wives of Korean men into the social and cultural spaces of South Korea (Kim, J. 2011). For participants in this research it almost invariably served as a blockage in attempts to imagine a future life in Seoul.

For many participants like Arturo whose embodied presence visibly distinguishes them from the ethnically defined notion of Korean personhood then, identity can be one of forever being foreign regardless of their own desires. Becoming migrant is not only about individual wishes and aspirations but also reflects encounters and consequent configurations of bodies, meanings and culture. Adit (Indonesia, Male, EPS worker), for example, spoke about how identification as a foreigner was reinforced in times of panic about irregular migration:

> At the end of my first five months work, there was a sweep for illegal workers. During that month, all the army, police, immigration and civil offices were collaborating to sweep the illegal workers. They will sweep all through the factories on the 11th and 12th.... It was so horrible. They checked each and everyone in the buses, train, and train stations. They cooperated with the drivers, checking if there was any foreign passengers. People who got into the taxi were not taken to their destination, but to immigration office.

At times when the migration regime manifests in policing of migrant bodies, identification as foreign amplifies one's sense of being out of place, of being present in the nation only at the will of government and people. It also reveals how the identities of migrants are shaped by governmental practice – the right to reside and remain – and their reinforcement by those who identify themselves as Koreans, the fact that train and taxi drivers would send all foreigners to immigration offices.

Foreignness is constituted in this context as a negative identity in relation to the 'reactive body' (Bignall 2008: 131) of Korean personhood. It is defined by its difference, and necessarily then contributes to the stabilisation of myths of ethnic homogeneity – one either is or is not Korean, and if not then necessarily foreign. Becoming foreigner then is often seen as beyond the control of mobile people themselves and is received by many as a matter of fact description of their legal status and identity in South Korea. Other participants who affiliated with South Korea also articulated a sense of difference and exclusion from possibilities of being Korean and consequently of remaining in Seoul. Two examples are useful here. The first is from Charlotte (USA, Female, English teacher), a Korean adoptee who described herself as being 'invisible' because she could pass as being Korean:

> I am invisible until I open my mouth and I start speaking in broken Korean; and so I would go out and I would sometimes, especially during vacation, I would just spend days without saying a word to anyone.... I selectively use my outer appearance to *pass* as Korean.

The capacity to pass as Korean means that, on an everyday basis, one is not identified by others as foreign but it does not automatically lead to the disruption of becoming foreign and having a heightened sense of being out of place. Charlotte also spoke about difficulties with language and cultural differences that meant that she could never consider herself Korean even though her migration was partly about connecting with her (unknown) biological parents' culture. Despite leaving the US with a sense of being Korean, she found that once she knew people personally in Seoul that she became necessarily foreign, a transformation that also involved becoming American in ways that were unanticipated:

> When I was in the US, I was Korean American no hyphen. I'm here now just here. I don't know if it's because I don't feel the need to distinguish ethnicity when Koreans are everywhere or just because coming here has made me feel more American. I've realised the ways in which I'm absolutely American. I don't know.

Huizhong (China, Female, Student SNU), who was also of Korean descent but was born in China, expressed a similar set of challenges around her identity, a struggle that emerged in encounters and their implications for considering herself Korean.

> They wouldn't say that I speak Korean very well, but most of them would say that, because I'm of Korean ancestry, therefore my Korean language should be of this level. ... I feel that point of view more keenly from foreigners. Foreigners will go 'oh, you're quite good ... to be able to achieve this standard is already quite good', but Koreans would go 'oh, you're of Korean ancestry, therefore you should be like this from the beginning'. ... In the past, back in China, I never had this kind of identity problems. ... I didn't have such a problem. But after coming to Korea, I don't know why this is so, but for the first two months, I was quite troubled. I found it very difficult to comprehend my identity.

Like Charlotte, then, language and recognition matter significantly in the way encounters contribute to processes of becoming foreign, or recognising oneself as having identity misalignments with Korean personhood. For her and others the sense of distance generated in encounters also contributed to desiring to be elsewhere and a blockage of any possibility of remaining in Seoul long-term:

> Korea? I think I want to leave the place where I experience clashes in identity. For instance, if I were to go to Europe or America, I think it shouldn't be a problem. But to stay in Korea, I think there will be a clash. ... Actually, I don't consider the place in Europe or America my home, but there I would be able to live more comfortably without such clashes of identity.

The expression of a 'clash' reflects what appears to be a fundamental incompatibility of Huizhong's sense of herself and her place as an ethnic Korean in South Korea with affiliations outside the nation (as Chinese). For her, remaining long-term is an impossibility because of this very problem of aligning oneself comfortably with people and place. While care is needed in recognising commonality between these identity crises and the repressive policing of foreignness that Adit spoke about what becomes clear here is the way in which becoming foreigner involves the arrangement of different entities: embodied differences, cultural expectations, migration regulation, language and senses of identity. These articulations of becoming foreigner in relation to Korean personhood are fundamental in shaping the subjectivity of migrants and the capacity to conceive of living in Seoul in the future. Put another way, certain kinds of encounters lead to a form of social closure, where even those with rights and capacity to live in Seoul recognise the difficulty of adapting or aligning themselves with expected norms.

7.2.2   *Alignments with Korean personhood*

Passing as Korean represents one albeit never complete way of becoming different through migration. It is something that is only possible for some people however – those of Korean descent and those who felt that they were identified as Korean in an embodied sense. Yanmei (China, Female, Student KU) who was Korean-Chinese, for instance, described how 'I met my best friend here and I feel I am one of them. I don't feel Koreans see me as an outsider since I feel I can fit in'; Emmanuel (Canada, Male, English teacher), whose parents were born in the Philippines, also insisted 'I don't feel out of place. I mean, I think I am just another guy in the crowd'; and Minsheng (China, Male, Student SNU) asserted 'we are all yellow-skinned, so if you were walking on the street, people would not be able to tell if you were Korean, Chinese or Japanese. You must go to the US [for that]'. As Charlotte and Huizhong revealed though, there are limits to these connections, moments when for reasons of language, culture or status that passing as Korean can sit alongside a feeling of being foreign. Becoming, however, cannot be understood as completely constrained by these predetermined identity categories of ethnicity. Indeed, we should understand becoming and its implications for migration as hinging on encounter: 'the reality of becoming is the reality of an encounter between affects of bodies that produces a new affect' (Marrati 2006: 320). We also need to look at the ways in which desiring makes it possible for migrants to imagine and enact alignments with Korean personhood:

> For the first time in Korea, it was really tough but after seven years, I feel like I don't think I am Vietnamese anymore. Although my Korean is not so good but I keep thinking that I am the same as other Korean people. I never think that I am a foreigner, a Vietnamese. Now I think I am a Korean, I live in Korea so I am a Korean ... I just think I am a Korean, always. Sometimes I don't care that I am Vietnamese.

In a few instances, like Thuy (Vietnam, Female, EPS worker), the time spent in Korea (seven years in her instance) has been a period where she has become increasingly comfortable with her position in society. She spoke about a feeling of not being a foreigner or Vietnamese, that she was 'the same as other Korean people'. It is important to emphasise that for Thuy this alignment with Korean personhood did not reflect an ability to pass as Korean because she noted how her neighbours asked if she was a foreigner and tried to find out about her life. Rather, after considerable struggle in her early life in Seoul Thuy had found a familiar and safe place in the city, where she knew her neighbours, had civil relations with her employer, and felt that she was accepted – she could imagine a life in Seoul into the future.

Time is a critical component of Thuy's process of aligning herself with various components of Korean personhood. Indeed, while she recognised early struggles it was the duration of her stay and the relations she had developed through this process that made it possible to configure not only an enjoyable present life but to project future possibilities in the city. Time also figured significantly in Pin's (Vietnam, Male, EPS worker) account of his sense of becoming at home in the city both now and into the future:

> Actually, I have lived in Korea for long, I don't feel strange anymore, it becomes something familiar for me.... In general, I am kind of playful person, so I feel satisfied with life in Seoul. During the time I live in Seoul, I nearly didn't go anywhere; I never went to other provinces. I just live in Seoul. Maybe I stay like that for long, I feel familiar with it because of my personality. I will keep living like this.

Pin did not articulate a sense of being Korean and no longer being Vietnamese in the way that Thuy did, but he did stress that he lived life as he wished in Seoul and felt a sense of familiarity and belonging in the city. It is worth noting that both Thuy and Pin were undocumented for most of their time in Seoul and South Korea; Thuy after leaving her brokered marriage and Pin after leaving his contracted employer. Rather than leading to long-term marginalisation, however, both Thuy and Pin stressed that it was time in the city that made it possible to craft an alternative sense of place – that they could be at home because of the seven and ten years that they had spent in South Korea. In contrast to the migration regime that would mark and enforce them as migrant workers whose time should be constrained, this longer period of time makes it possible to develop new and unanticipated alignments with place, to acquire language, skills, friendships, professional reputation and familiarity that makes it possible to imagine life beyond the constraints of their prescribed identities.

If establishing alignments with various aspects of Korean personhood or life in South Korea takes time, then it also demands a capacity and desire to align oneself in specific encounters. Indeed, as some examples from English teachers revealed, tenure in South Korea does not lead to closeness to or affiliation with Korean personhood. There were teachers who had been in Seoul or South Korea for more than a decade but had little interaction with anyone beyond the English teaching community; longer durations may also provide scope to construct parallel and relatively immutable worlds of culture and identity. As such, because of its very stable and clearly demarcated contours, aligning oneself with Korean personhood demands effort – learning language or adopting behavioural traits that smooth encounters and relationships. Siri (South Africa, Female, English teacher) explains:

> I had another friend, also white South African girl and I got into a discussion with her about how Koreans treat me, and have opened their homes to me. I'm being introduced to grandmothers, families, and she said to me, 'That's not common for them to be so open to a foreigner'. Difference is, also, I will tell you more, I'm of Indian descent, Asian, and a lot of things that I do, gestures to my co-workers. They tell me you're just like *Hanguk Saram* [Korean person], the only thing is your skin is a bit darker, but my manners, my behaviour, the way I treat others, is you know.... And I often take many things to school like food, or something from the bakery or something for the teachers and this South African that I told this to, she was like 'why?' Just because they're very good to me and that's just my way of showing that I like them and you know, and that's what teachers do. They take goodies to school, then share out.

Siri's direct comparison of her seemingly innate connections with Koreans and her 'white' friends' disbelief reveals some of the subtlety of aligning identities and practices. In this instance, alignments with Korean personhood appear to result from recognition of similarity in bodily comportment – gestures, manners, behaviours, deference to others – all common norms that serve as part of the constitution of identities. The notion that Siri might also be 'just like *Hanguk Saram*', however, also shows both how she is introducing difference into these identities and how the Koreans she interacts with are willing to see flexibility and mutability around the edges of identities.

Becoming aligned with established subject positions like being Korean is not simply a matter of working on the boundaries of inclusion. Rather, becoming aligned with Korean identities can expose individuals to efforts at reterritorialisation that may also police their own presence. Melanie (UK, Female, English teacher), who was married to a Korean man and worked in a regular company rather than in the foreigner-scripted spaces of English language academies, explains:

> I cannot behave in a normal Western way in the office. Impossible. I have to make extra efforts to adhere, accepting their authority, no challenging whatsoever. I can't even make a suggestion about workflow 'cause I'm being difficult....The better Korean you speak, the worse that becomes. Really. If you do not speak Korean, they assume you're just a foreigner. If you speak Korean and you're married to a Korean, then 'Oh [Melanie], you've lived in Korea this long, you should understand this' and I go 'No' you can't say that.

Disrupting the boundaries of what it means to be Korean through language abilities, employment, time in Korea or marriage to a Korean is not something that is only directed by the agentive will of those on the margins of this identity.

As Melanie demonstrates, there is a struggle between others' insistence that she aligns herself more fully with being Korean and her own drive to hold herself at a distance. This is also a gendered experience of becoming included in social norms – none of the male English teachers married to or in long-term relationships with Korean women or men discussed expectations to conform to notions of Korean personhood. In contrast, Martina's account and Mary's discussion in Chapter 6 of the advantages of being married to a Korean man reveal an overlap with discourses of multiculturalism that favour the inclusion of marriage migrants as wives and mothers over other kinds of migrants.

7.2.3   Coupling and decoupling futures

The sense of becoming and remaining a 'foreigner' and, by contrast, of opportunities to work at the edges of Korean personhood are important features of the emergent diversity that remain unacknowledged in the politics of urban multiculturalism in Seoul. These different socio-cultural encounters and becoming of migrants speak to the challenges of new forms of migration that are seen as necessary in an economic and demographic sense and desirable in the crafting of a *Global Asian City*. They reveal that multicultural presence can also be tied to fractured futures where the desire and capacity for individuals to plan for and imagine a life in the city varies considerably. There is, in other words, a coupling and decoupling of different migrant and urban futures. In this last section I draw on three examples of individuals in this research projecting into the future as a means to identify some of the many connections between desiring-migration, urban encounters and prospects for the future. Each person had been in South Korea for many years and were reflecting on the possibilities that the future held and their ongoing connections with Seoul.

Pin (Vietnam, Male, EPS worker), as we have already seen in Section 7.3 and in Chapter 4, had led a remarkable life in Seoul. He had arrived during the last years of the industrial trainee scheme, became undocumented and remained so for nearly a decade during which time he established a highly successful work life where he was well compensated and empowered in the workplace in important ways. He articulated a comfort in life in Seoul that was uncommon across all participants in this research: 'I am a foreigner and live illegally but I don't feel I am under threat, or deprived of the human rights. In general, I feel so comfortable'. He is a clear example of the way in which migrants exceed what is expected of them by the state and appropriate a space in the city for themselves despite efforts to restrict this.

Despite his assertions about 'wanting to keep living like this', by the end of his interview Pin also acknowledged the inevitability of his eventual departure from South Korea. There were several reasons for this, including a desire to reconnect with his family after many years' separation but also a recognition that 'some time

I will have to leave' because of the impossibility of becoming legalised. Departure in Pin's case appears to be final. If he leaves as an undocumented migrant he is very unlikely to be granted the right to return to South Korea in the future. In this regard, his circumstances highlight the regulatory force of the migration regime to restrict the present lives of migrants and their future place in the city and nation. So long as he is willing to subvert this regime Pin has developed the capacity to remain in a seemingly indefinite way but once he desires more than this peripheral existence departure appears like the only possibility. There is then considerable unfreedom in migration even for those who demonstrate agentive will and capacity to the extent that Pin does. It was in this context of temporal closure that Pin started reflecting seriously about what comes next:

> Before I go back to Vietnam I have to investigate some issues which will be a foundation for me at home. I mean when I go back home I want to do everything related to Korea because after many years I have been living there, I got to know Korean language and gain a lot of experience. At the moment I have started a forum called Korea consulting forum. I want this forum to become a bridge to provide information about Korea for people who care, a place for people to share experience about life. Because I think that the Vietnamese have a long history of coming to Korea but there is not extensive knowledge sharing among people, so the newcomers have to search and get to know Korea while the old ones don't raise their voice.

This set of plans and their intended effects point to the ongoing possibilities of migration and the way that individuals like Pin capitalise on their experiences not just during their time constrained period in Seoul but through the future oriented assembly of life. They reflect how he has been drawn into worlds of opportunity not only for his own benefit but also to make further migration possible. While his departure was imminent, his future plans point to becoming increasingly central in the migration assemblage (Rubinov 2014), about being one of the gears or cogs in the machine that makes migration possible; desire, in this context, can be understood as a person's 'fascination with these gears, his desire to make certain of these gears go into operation, to be himself one of these gears' (Deleuze & Guattari 1986: 56). Many migrants remain at the edges of these arrangements, undertaking migration through relations to actors, materials, ideas and systems but not becoming central to the operation of these assemblages. Others like Pin reveal the force of desire not only in the specific practice of migration but also becoming otherwise in that process in a way that alters subjectivity and place as well as the configurations of migration itself.

With her graduation approaching the following year, Yunru (China, Female, Student SNU) also reflected extensively on future prospects and her place in the world. She was completing a PhD in Korean language and was working in translation for a Korean firm operating in China while teaching Mandarin as a part-time job. Like Pin, she was fluent in Korean and comfortable in

Seoul – reflecting that 'I have a better understanding of Korean culture and have become more accepting of things'. Yunru was one of a smaller number of international students in this research who were able to align themselves with Korean personhood. In her case this involved building relationships both amongst other Chinese students but also Korean students and her supervisors in a way that supported feelings of belonging in Seoul.

Yunru placed a significant emphasis on her age and also on a desire for 'freedom', an ability to go where she wishes and pursue opportunities as they emerge. She was 30 years old at the time of the interview and described how she was 'very anxious, I would like to graduate as soon as possible'. She wanted to avoid 'just floating along' and rather move on to the next stage in her life so she could 'go on to do other things'. The timing of life here reflected a sense of progression but one that was both gendered and tied to familial responsibilities. It was being a *woman* at 30, without a career and unmarried that impeded her progression:

> I'm very anxious. I'm already 30 but I feel that a woman isn't a woman already; everybody at every stage of life has some things that they should do. If I were only 27 years old, I wouldn't feel anxious but would feel that I'm in a good state. But because after 30, I feel that I should at least have a job and a family but I have neither, therefore I feel very anxious. ... I'm really worried about this, my parents have been supporting me till this age, I cannot continue stretching out my hand to ask them for money, it is very tough for them. At my age, I should be supporting them, should be giving them a good life.

Becoming woman here is also part of the migration process, completing study, finding a job and securing a stable domestic relationship conforms to expectations of life course progression that circulate widely but are in this case specific also to China (Fong 2011). The notion of filial piety is also significant here in orienting Yunru, the obligation to support parents after their investment in her study and life. Yet, Yunru's views are also shaped by a desire for freedom and mobility, she did not want to feel like 'I'm a person made of paper, fixed in one place and not able to move'. Instead, she spoke of pursuing transnational careers:

> I will follow my work. That is to say, wherever my work takes me, I'll go. If my work requires me to stay in Korea, I don't mind doing so. But I don't have a strong desire to become a Korean citizen or that I want to continue staying in Korea. I'm not very picky about the location where I am at. I like to be mobile. My ideal would be to have a home in [my hometown], but I can work in different places in the world. I can work for a few years in this place and a few years in that place.

The time spent studying at SNU and living in Seoul here serves as a stepping stone or staging point for enhanced mobility post-graduation. The experiences she has had, the credentials she has gained and the language and cultural competencies acquired through encountering life in Seoul generate a significant capacity to move, and a personal orientation that frames this as desirable. There are similarities with Pin here in that Yunru is looking to capitalise on her time in Seoul but she also clearly feels the freedom to pursue futures in the city, to travel to other locations or to return home if and when that suits her. Migration here is generative of desire for further migration but her capacity to pursue her desires is also supported by the migration regime that privileges her presence as a future talent migrant. Such 'freedoms', as Yunru put it, are a product of the liberal control of the migration of globally educated people in South Korea, but it is also about her own investment, 'the making and remaking of one's own life' (Papadopoulos & Tsianos 2008) in the present and into future possibilities.

Mary (Canada, Female, English teacher) initially migrated to South Korea in 1997, worked in Gwangju for a year, spent two years in Canada before returning in 1999 to work in Ulsan, Masan and Daejeon. She had been married and had one daughter with a Korean man but was now divorced. She was working at a major university at the time of the interview, felt that she was 'communicatively competent' in Korean but not quite fluent, and had a range of friendships with Korean colleagues, some of the more established English teachers and other migrant communities. Locally too, she emphasised that 'I really like feeling like I'm part of the neighbourhood ... I love living here'.

Mary's future possibilities were shaped by this sense of belonging, by her own recognition that 'my whole adult life' since 22 years old was based in South Korea. She also noted that when she visited family in Canada every two years that 'I don't really feel Canadian'. More than Pin and Yunru then, Mary was very attached and connected to her life in Seoul – she imagined the possibility of remaining but was now confronting the implications that held for herself but also for her three-year-old daughter:

> I really don't know. I mean I've never, I really ... a really important time I think in my ... life. I'm going to do my PhD but I don't know if I'm going to do it in the UK which would be great but really expensive and I don't know how practical that is since I'm Canadian. Or I could do it in Canada which would essentially mean I'm telling myself and [Harriett] that we're settling in Canada at this stage or in Korea I could probably do it here at SNU but by doing that I would be making a decision to stay in Korea basically. So that's my future.

The timing of lives is a significant influence on how Mary sees herself in the present and the future. The time that Mary is reflecting on here is not only her own but also the time of her daughter, her own progression through to the Korean education system.

> I think about [Harriett], she's Korean and Canadian so when I make these big decisions about life and work and where we're going to settle, of course I'm making decisions for her. So, I really want her to speak Korean; she does actually. I want her to keep speaking Korean, I want her to know she's Korean. But at the same time,

I'm not sure if I want her to stay here and go to school here. . . . Even though living in Korea is really hard sometimes as a foreigner I think I'm still more comfortable here because I've lived here for so long so it's a big decision to go the opposite way – to pack my bags and take all my stuff and my daughter and move back to Canada. So, I worry that I'm making the decision because it's really a non-decision.

Migration in this account, but also in many ways in Pin's and Yunru's and many other participants' reflections, is not only an individual pursuit framed around goals that can be clearly determined in advance. Rather, it involves anxiety about the future, love for those we are in relationships with and complex and sometimes conflicting senses of belonging to the places or people migrants encounter. Desire emerges in such moments, not as clearly demarcated and obvious objectives, but rather as multiple forces drawing individuals in different directions, highlighting the power of different configurations that create opportunities or potentially block mobility. In these reflections, the configuration of Seoul as a place only for a temporary even if extended stay becomes amplified. In the end Mary articulated her eventual but still not determined departure as 'a non-decision', something framed by the reality that while she can sustain a life in the present the future possibilities for herself and her daughter appear oriented elsewhere.

7.3   Conclusion

The discourses of urban diversity and circulation articulated by Lee Myung-bak and Oh Se-hoon, as well as the framing of an 'irreversible' movement towards Korean multiculturalism, are all responses to the growing significance of migration in Seoul and South Korea. They position migration as a challenge but one that brings positive externalities if it is appropriately managed; the orientation of the city and nation can be reconfigured to incorporate those migrants who are deemed desirable. Alongside the wider regulation of migration that has been discussed throughout this book, these efforts to reconfigure the city and nation towards globalising processes involve considerable segmentation in the present as well as the future possibilities for migrant lives. This reflects Aihwa Ong's (2007: 92) view that:

> While the cosmopolitan mirage of megacities project a multicultural globality, the urban condition is shaped by divisibility and even incommensurability of human worthiness rather than by a fusion of multicultural horizons that consolidates our common humanity.

As I have shown in this chapter we can see this divisibility in the everyday urban encounters and modes of social inhabitance that migrants engage in, the subjective alignments of migrants with embodiments of foreignness and Koreanness and in the future prospects that migrants have in Seoul. First, as I have demonstrated in discussing the social practices of migrants in the city, urban multiculturalism does not take shape in an additive fashion where migrants serve only as symbolic capital for the aspirant global city. Rather, difference emerges in articulation with urban spaces and the different ways in which desiring forms of migration and everyday life take shape in encounters with people, objects and ideas. There are already emergent forms of multiculturalism or urban diversity expressed and manifested in places like Itaewon, Hyehwa and Ansan that reveal Seoul's extant diversity. Often such spaces emerge through desiring familiarity and stability on the part of migrants and as a result can take shape around the representation of teachers, workers and to a lesser degree students as distinct groups of migrants. In other instances, there is aversion or avoidance evident in the narratives of migrants, desires for other kinds of relationships with people and place that take shape in less visible but perhaps also significant micro spaces of social life in the city. They open onto other possibilities that are not beyond the migration regime but perhaps rework or even at times exceed its expectations. The divisibility that Ong speaks of is evident in these manifestations then, but there is also clearly a sense of possibility expressed that lives could be oriented differently and that urban diversity could take shape in more open encounters.

In addition to observing this generation of difference in urban space this chapter has also turned to the diverse ways in which migrants themselves articulate a sense of transformation in subjectivity. There is a pervading sense of becoming foreign through coming to Seoul, where most participants, even those with Korean ethnicity, struggled to align their own identities and embodiments with notions of Korean personhood. These encounters remind us of the significance of the myths of ethnic homogeneity that still structure national identity in South Korea and shape possibilities of migrant inclusion. And yet, there are instances where individuals from a range of backgrounds do make connections with Koreans, and do recognise themselves as becoming somewhat Korean. While this book does not extend to a detailed analysis of what this means for Korean multiculturalism, it does point towards the untenability of maintaining myths of immutable ethnic difference in the context heightened migration.

Last, the chapter has shown how those who have become most accustomed to life in Seoul, who have spent many years there, also need to figure their own futures. Even in these cases of strong familiarity with place, familial connections and a desire to remain, there exists anxiety at least if not regulatory or socio-cultural blockage on a future in Seoul. While few migrants in this research desired such a long-term future in Seoul, the foreclosure of this possibility nonetheless gives pause for thought about the manner that migration futures are decoupled with those of the city. To return to Lee Myung-bak, when we 'think of the future of Seoul as that of a "world-class city"' we are forced to recognise an urban future imagined at once as globally connected and open and yet clearly reliant on uneven and temporally constrained inclusion of the migrants who are all critical to enabling this globality.

Acknowledgements

Portions of this chapter have been drawn from Collins, F.L. (2016). Migration, the urban periphery, and the politics of migrant lives. *Antipode*, 48(5), 1167–1186. doi: 10.1111/anti.12255 and Collins, F.L. (2016). Labour and life in the global Asian city: the discrepant mobilities of migrant workers and English teachers in Seoul. *Journal of Ethnic & Migration Studies*, 42(14), 2309–2327.

Chapter Eight
Conclusion

8.1 Global Asian City

The development and transformation of Seoul has often been historically framed as a fundamentally indigenous achievement, one that reflects the unity, resilience and capacity of the 'Korean people' (Kim & Choe 1997). Setting aside the rupture of invasion, occupation and colonisation by the Japanese, the South Korean capital has been narrated as a key node in historical processes of feudal consolidation (Yi, T. 1995), national modernisation and industrialisation (Cho 1997) and a basing point for global expansion (Watson 2013). These framings of Seoul, which easily lead to its popular moniker 'the Seoul Republic', speak to a nested geography wherein the city serves as a key unit in national development and wherein urbanisation results from fundamentally internal or at least nationally bordered processes. In the case of Seoul, this has also meant that processes of migration have not been considered a key part of the making of the city. Indeed, as recently as a decade ago, scholars would propose to examine 'the *absence* of labor migration' in the making of 'Global City-ness in Seoul' (Kim, E.H. 2008: 323, emphasis in original).

In this book, I have taken a different approach to conceiving the city of Seoul, and more broadly the relationship between migration and processes of urban transformation and consolidation. In particular, I have suggested at the outset that to examine urban life in Seoul today it is not sufficient to address only those people, things and practices that are coded as internal and indigenous to the city. Rather, we also need to include in our analysis the migration, arrival and lives of

Global Asian City: Migration, Desire and the Politics of Encounter in 21st Century Seoul, First Edition. Francis L. Collins.
© 2018 John Wiley & Sons Ltd. Published 2018 by John Wiley & Sons Ltd.

migrants in the city and the possibilities that emerge through what have often seemed like unexpected interactions between people and place. As the accounts of different migrants in this book have revealed, migration does not only occur on the relatively stable backdrop of the city and nation but is rather entangled in reconfigurations of those geographies that alter both present and future possibilities. Migration occurs in relation to changing imaginative geographies of the city and nation, is connected to political–economic shifts as well as new approaches to regulating the flow of people across borders and the planning and management of urban spaces. Migrant lives are situated into relation to these and other dynamics, they do not result only from rational decisions and regulated openness but rather also in impulses and movements towards the unknown and in the unpredictable encounters that constitute urban life in Seoul and cities around the world.

In this regard, the arguments that I have developed in *Global Asian City* are situated most notably within a growing scholarly interest in the relationship between migration and cities (Collins 2012; Glick Schiller & Çağlar 2011; Ley 2010; Ye 2016a). Most importantly, I have sought to move beyond established *migration-centric* accounts that treat cities as self-evident backdrops to migrant lives and *urban-centric* accounts that view migration as a relatively common sense process of populating cities. More recent attempts to explore migration and cities have advanced understanding by focusing on cities as 'gateways' and 'entry points' to wider geographies (Price & Benton-Short 2008), on the role of migration in urban restructuring and its effects in the built environment (Ley 2010; Mitchell 2004) and on addressing the varying 'pathways of incorporation' that migrants take in urban life (Glick Schiller & Çağlar 2009; 2011). Theoretically, however, these approaches have not developed a conceptualisation of *both* migration *and* urbanisation that might link the drivers of migration to the transformation of cities, the different positions migrants hold in society and the contested politics of everyday life. In part, this shortcoming reflects a continued focus on western immigrant cities as a site for theory making wherein migration has been understood as a pre-existing and relatively expected set of processes. It has been more challenging for scholars to conceive of the initial generation of migration in these contexts and the varying ways in which that might be related to recent or more established urban transformations. Accordingly, the drivers of migration have often been treated as obvious (economic advancement, lifestyle, transnational social networks, settlement and citizenship), and there has rarely been a close focus on the differentiated character of migration and its articulation with urban processes.

Seoul has represented an ideal case to unpack emerging entanglements of migration and metropolitan transformation. Indeed, as the chapters in this book have demonstrated migration has over the past three decades shifted from a relatively invisible phenomenon at the edges of South Korean society to one of the most pressing social and political issues, particularly in Seoul and its wider urban region. International migration represents a transformation that has emerged in relation to the consolidation of the nation and its borders and the extension of the city of Seoul into regional and global territories. Seoul, constituted in its built form and social life but also through an increasingly mobile Korean population, the economic activities of Korean firms, the production and circulation of popular culture and the restructuring of higher education institutions has become interlinked into new patterns of migration that are opening the city to a wider range of flows and encounters. This is apparent in the way that changes in the geography and technology of manufacturing and the expectations of young Koreans are connected to the recruitment of workers from outside South Korea; in the increasing importance of English as a sign of global status for individuals and families, and at urban and national scales; and in the emphasis placed upon international students as symbols of globalisation in higher education and future human capital for demographic renewal. In each of these cases, migration has not simply occurred *to* Seoul but rather has been embedded in other transformations and has led to alterations in the structure and experience of urban life from the periphery, to suburbs, factories, schools, universities and public spaces.

Migration, however, does not simply just occur and neither can it be understood to be only the result of rational decision-making and future planning. Rather, the focus on the *emergence* of Seoul as a place constituted through migration has also demonstrated the importance of conceiving of the way in which migration must be generated and the wide range of forces and actors involved in this. Indeed, the diverse narratives of migration discussed in this book do not speak to long established ideas of migration as a natural undertaking or to Seoul and South Korea as inevitable and unsurprising destinations. Rather, migration is something that must be generated through encounters with the possibilities that appear to manifest in notions of moving in the world and the characteristics of places. At times these possibilities have been connected to the different undertakings or occupations of the women and men in this book – of completing a qualification that is understood as holding value, earning money to pay back debts or to support family life or creating opportunities for future careers. Throughout the chapters, however, it has been the imaginative dimensions of migration that have been particularly important – in triggering ideas about what is possible through migration as well as in directing individuals towards Seoul and South Korea as places for achieving certain kinds of things. As we have seen too, such imaginations are never fully formed in advance but rather emerge through encounters with different kinds of intermediation: the work of profit-seeking brokers and agents; the regulatory apparatus of the South Korean state; the stories told by friends and family; the recruitment activities of institutions and employers; the circulation of imagery through popular culture; and the vast and dynamic imagery and information that circulates online. These intermediated imaginative geographies have been configured in ways that work through the specific circumstances of the different individuals and groups in this book, that cultivate and target different desires for becoming mobile and work on embodied impulses for addressing present concerns through migration. Contemporaneously, these imaginative geographies have also reflected ideas about how Seoul and South Korea have transformed since the Korean War and in doing so have spoken to exactly the possibilities that exist as a migrant – worker, teacher or student – in Seoul in the twenty-first century.

The relationship between migration and cities as it has been discussed here has also highlighted the territoriality of the nation state and the establishment and management of border technologies for regulating flows of people. Indeed, while we can view the transformation of Seoul over recent decades as fundamental to the generation and implications of migration, the various movements discussed in this book have also been permitted and managed in ways that shape their urban dimensions. One of the key characteristics of the South Korean migration regime and others like it in Asia (Hoang 2017; Ye 2016a; Zhang, Lu & Yeoh 2015) has been the manner in which it has treated the migrants discussed in this book differently in ways that reflect presumptions about education and skills that also link into notions of nationality, ethnicity, gender and age. Accordingly, while international students can arrive from any nation so long as they qualify for pro-grammes and can arrange for the payment of fees, workers and teachers must qualify both in terms of skills and education as well as nationality and more subtly gender, ethnicity and age. Only citizens of seven Anglophone countries may become English teachers and only citizens from 15 Asian nations can enter South Korea as labour migrants through the EPS. Both groups require particular educational qualifications and for workers must pass assessments of Korean lan-guage; teachers and workers also need medical and police checks and teachers are tested for HIV and drugs. More intricately, EPS workers are overwhelmingly male, a pattern that does not reflect formal regulation but rather the gendering of particular occupations in both migrants' home countries and South Korea as well as the expectations of employers and the way in which this folds into recruitment and selection processes. Age too is important, both for EPS workers who need to present themselves as able bodied but also for English teachers who are expected to be relatively young and recent graduates.

These formal and informal 'hierarchies of regulation' (Lindquist, Xiang & Yeoh 2012) differentiate who can enter and what individuals can do within the *nation* but as I have shown in this book they also fundamentally shape urban lives and the manner in which migration alters the city. The analysis of the migration of these groups reveals the substantial unevenness to their spatialisation in Seoul. Rather than a backdrop or context for migratory processes, the urban landscape is an active constituent of the differential incorporation of mobile subjects. There is a regulatory gradient to these differences, with migrant workers inhabiting the urban periphery, English teachers living and working across the urban region and international students positioned in the urban core. The state plays a significant role here in setting the conditions of different visa types and the relations between migrant workers and English teachers and their employers, and to a lesser degree between students and universities. At the same time these spatialities also demonstrate broader political–economic transformations in Seoul, where most manufacturing industries have increasingly shifted to the outskirts of the region, English language academies have become a normal part of residential areas and the most internationally oriented universities have emerged in the centre of Seoul. As the accounts of migrants in this book have revealed, these different urban spaces offer varying opportunities to participate in urban life – either in terms of being visible in public spaces, interacting with local populations and other migrants or contributing to longer-term changes in the spaces of the city.

Contemporaneously, migrants inhabit urban space in ways that relate to the multiple and differing temporalities of their lives (Robertson 2014). Again the migration regime looms large here, with the temporal limits of visa categories and the possibility or desirability of a long-term future in the city varying significantly. While there appears to be relative openness to international students remaining in Seoul and becoming graduates and then skilled workers in a regionally connected economy, the lives of English teachers and migrant workers are subjected in different ways to temporal constraints. English teachers appear confined to an imminent departure or alternatively to its cyclical postponement in another 12-month contract that extends their stay but does not alter their socio-legal position. By contrast, migrant workers are subject to tight disciplining of temporal limits – either depart after a fixed duration or face the precarity of undocumented life and its implications for their place in the city.

These different temporalities of migration also articulate in important ways with the everyday lives of migrants. Each group is subject to differing daily rhythms that are to varying degrees imposed upon them: from the regimentation of the duration and intensity of the factory floor, to filling time between English classes, to university schedules that assume attendance but expect self-monitoring. Beyond work and study, time also shapes engagements with public life in the city. Both migrant workers and English teachers appropriate parts of the city at certain days and times of the week, generating transitory sites of attachment that allow for a separation between work and social times. Despite their transience these appropriations of public space are significant because they allow for the construction of alternative socio-cultural modes of life that allow migrants to be more than just labouring and learning bodies. They are about staking a claim to being a public in the city. These spaces, though, are also configured in relation to the consolidation of identities, and can involve racialisations, gendered norms and lifestyles that can be familiar and stable to some while generating aversion and avoidance amongst individuals seeking different kinds of encounters. The urban is produced through these and other temporalities, an urban-region that is always taking place, that is brought into presence through shifting combinations of bodies, regulations, habits and relationships (Simone 2010).

It is in figuring these entanglements of urban transformation, national regulation and migrant lives that the dynamic relationship between migration and cities
can be brought to the fore. In *Global Asian City* this has been achieved through
empirical work that brings together migrant subjects who are not normally conceived in relation to each other as well as through a distinctive theoretical approach
that links ideas of desire and assemblage with a focus on the encounters that constitute urban life. This framework has made it possible to show how the generation of migration (desire) occurs across the migration and lives of migrants, is
shaped by shifting urban forms and national regulation (assemblage) and has
consequent impacts on the everyday lives and *encounters* in the city. In addition to
shedding light on the specificities of the interaction between urbanisation and
migration in Seoul over recent decades, this theoretical approach also provides
three wider contributions that I detail through the remainder of this chapter.

8.2 Desire, for Another Ontology of Migration

There remains within mainstream migration studies a reliance, even if now largely
implicit, on an understanding of migrants as rational calculating subjects who
knowingly use migration as a mechanism for economically enhancing livelihood
(Nail 2015). This view not only dominates scholarly accounts (De Haas 2011)
but also infuses the logics of migration management regimes that pursue the ideal
of ordered and regulated flows (Ghosh 2007). By introducing a focus on *desire* as
a social force the discussions presented here in *Global Asian City* have provided
an alternative conceptual or ontological basis for understanding migration that
links the generation of movement to its mediation and reconfiguration as migrants
make their way through the world. As the narratives of migrants presented here
have shown, migration is not guided by singular pursuits of economic accumulation
but is rather also caught up in imaginative, embodied and emotional drivers. The
seemingly very different narratives of workers, teachers and students have demonstrated the unpredictability and excess that is manifest in migration. Migration
involves migrants both being directed and directing their own mobilities, is situated within widely dispersed social relations and is configured within imaginations that encourage, enlist, channel or impede movement within the world.
Desire as it has been used here is not then simply a synonym to replace atomised
notions of 'choice', 'decision', 'motivation' or 'want' in migration studies. It
should not be used as a term to describe things that individuals consciously set as
objectives, but rather as the very instigation of becoming mobile in the world.

As a means of reconfiguring scholarly understandings of migration, a focus on
desire as a social force offers at least three significant advantages that have been
revealed in this book. First, while desire is very much a force that animates life in
manifold ways, we have seen how migration relies on the generation and channelling of desire towards particular ends. Migration generally, let alone migration to

Seoul and South Korea specifically, has clearly not been inevitable in the lives of any of the migrants in this research. Rather, their diverse movements have occurred in relation to encounters with imaginaries about what is possible through migration, in blockage from seemingly more attractive possibilities and in the channelling of desiring-migration by particular migration regimes. The imaginative dimensions of migration are all too often taken for granted by migration scholars, a tendency that then serves to reduce migration to assumptions about economic calculation and advantage (Carling 2002; Carling & Schewel 2017). This book has shown that migration is not simply the result of a decision about a place that can be known a priori of its representation and mediation but rather is generated in embodied encounters with imagination and potential that play a role in constituting that very place.

Second, the focus on desire as a social force also highlights the multiple temporalities of migration, the manner in which the generation and effects of migration are distributed through migrants' lifetimes (see also Collins & Shubin 2015; Griffiths 2014; Shubin 2015). Initial encounters with the possibility of Seoul and South Korea and of migration more generally do not result from a singular moment of 'decision-making' but should be conceived across past, present and future temporalities. The idea of migration emerges in ways that are connected with individual and familial situations and biographies, in terms of imaginings of places, other people's migration as well as the intermediation of a wide range of social actors. Moreover, while migration is in many instances future oriented it is critical to also recognise how opportunism, the time of grasping the moment, can intersect with and be narrated around more distant and less determined temporal horizons. Throughout this book we have also seen how the generators of movement are altered and reconfigured in relation to the spaces and opportunities encountered by migrants. There is often a kind of 'anticipatory politics' (Simone 2010) at work in migration, taking chances as they arise, speculating on opportunities or taking risks in order to alter positions or to achieve things. In such moments, the temporalities and directionality of migration can be reworked and prior objectives might be reconfigured to suit new circumstances; migration is necessarily nonlinear in that it cannot be known and acted upon in completely determined ways in advance. Even in the future plans of migrants we see shifting notions of desire that are reworked through migration and encounter in the city; we see desire as the very remaking of one's life on the stage of the world (Papadopoulos & Tsianos 2008).

Third, understanding migration through the conceptual vocabulary of desire draws attention to the ways in which migration necessarily involves becoming as much as being (Bignall 2008; Grossberg 1993; Papadopoulos & Tsianos 2008). As the narratives of migrants in this book have shown, their imaginaries, migrations and everyday lives do not simply conform to the expectations of the migration regime. Rather, migration involves considerable agentive will on the part of migrants, they subvert regulation, become key actors in organisations, insist on their presence and belonging and rework identities to align themselves with being Korean rather than being marked as forever foreign. In doing so migrants can be seen to both appropriate and rework the territorialising powers of migration regimes – they become the labouring and learning bodies desired in these regimes as well as active human subjects whose presence cannot be completely contained (Mezzadra & Neilson 2013). The focus on desire reveals that becoming in migration is more than the result of calculative thought on the part of these migrants seeking to intentionally alter their circumstances or only the result of state intervention in the management of migration. Rather, it demonstrates the way in which migration is always about transformations of subjecthood, about becoming more than just a migrant.

8.3   Migration, Desire and Urban Assemblages

The discussions of migration and the city in this book have also offered scope to extend recent efforts to conceive of cities as urban assemblages in both empirical and theoretical terms. While accounts of assemblage urbanism (Edensor 2011; Farias & Bender 2012; McFarlane 2011a, 2011b) and postcolonial readings of cityness (Nuttall & Mbembe 2008; Simone 2010) have been open to migration, they have not to date explicitly explored the role migration plays in the constitution of urban life. By drawing attention to the importance of migration in the transnational extension of urban connections in a city like Seoul and their territorialisation in everyday life it becomes possible to account for some of the forces involved in processes of assembly without resorting to larger structural influences.

Addressing migration more specifically in accounts of urban assemblages provides another perspective on processes of assembling, not only through the active construction and inhabitation of built form in cities (Edensor 2011; McFarlane 2011a) but also through the literal peopling of urban assemblages. As we have seen in this book, the territorialisation of Seoul in the contemporary moment must be understood in relation to the movement and control of people. The peopling of cities, however, is not simply about the addition of new bodies to the assemblage, about an expanding size and scale. Rather, following Lee O-Young (1994: 72), we must observe that 'when [the city's] contents change, its shape [and character] naturally changes as well'. While the internal migration that helped to constitute twentieth-century Seoul made it possible to name and express the city as a fundamentally Korean achievement, the more recent migrations that have been the focus of this book are having a deterritorialising effect on this image. New flows of migrants have altered experiences of urban life in Seoul in ways that have demanded a new focus on urban diversity and national multiculturalism that seek to reterritorialise the emergent global city into a more stable form. Put in the language of assemblage, we can recognise that it is not only the properties of the elements that constitute the assemblage (including migrants and other urban residents) but also the manner in which they interact and are ordered in relation to each other (DeLanda 2006; Grosz 1994).

A focus on migration also reveals in important ways how urban assemblages involve diverse components and effects, they act on 'semiotic flows, materials flows and social flows simultaneously' (Deleuze & Guattari, 1987: 23). In this book, for example, we have seen how Seoul is not only constituted as a territorially embedded assemblage but also circulates widely through images of the city (semiotic flows) that draw potential migrants towards what is possible in migration. This is particularly apparent in the manner that the imaginative geographies of an advanced and desirable Seoul/South Korea circulate widely in Asia to generate the migrations of workers and students, while in the case of English teachers these imaginative geographies typically subsume Seoul and South Korea into wider imaginings of Asia or the East. These expressions of desire matter not only for generating migration, but also for then shaping how individuals become part of urban assemblages in their daily lives – their perception of what Seoul should be like, their orientation to individuals they encounter and their visions for their own futures. These flows are part of extending and reterritorialising the urban assemblage both spatially and temporally, they provide for a wide range of actual and virtual encounters as well as drawing new material and social flows into Seoul.

In drawing attention to migratory flows and their articulation into relation and co-functioning with Seoul, this book has also revealed the heterogeneous elements that are involved in the process of assembling the urban. Most notably, we have seen the role of deterritorialising and reterritorialising movements on the part of the nation state to open up and also control the movement of people. Urban life in Seoul is influenced by migration policy, especially for migrants but also for those places that are constituted through these flows. However, efforts to control mobility are also intermediated in the relations they establish with migration: migration agents and other governments cooperate, alter or impede the intent of South Korean migration policy to varying degrees; social networks or 'mobile commons' provide avenues both for potential migrants to encounter the possibility of Seoul and also to find alternative pathways both to and through the urban assemblage; and encounters in daily life generate both expected and unexpected transformation in the place and potential that migrants have as part of the city. It is at the interstices of these elements that we also see very clearly the ways in which the urban assemblage is organised and stratified (Grosz 1994), not in a straightforward top-down manner but rather in a way that reflects an ongoing play of the actual and the possible (McFarlane 2011b).

In all of these instances desire is critical to both examining and acting upon assemblages. We cannot conceive of an assemblage without the social forces that generate it, without paying attention to the production and channelling of desires. In this respect, I would argue that a key contribution of addressing migration within urban assemblages *and* drawing on notions of desire is that we begin to grapple with the explanatory problem that has haunted recent work on assemblage thinking

(Brenner, Madden & Wachsmuth 2011). Desire as it has been developed in this book makes it possible to examine what brings different heterogeneous elements together through the process of migration as well as their capacity to interact, create new forms and move apart from each other. The migration and lives of the migrants discussed in this book are exemplary of the effects of desire as a social force and the way in which it actually creates connections with Seoul and South Korea, imaginative, material and embodied, while also then constituting an ordering and stratification of everyday life and future possibility. While the focus here is necessarily partial given my emphasis principally on migration, this is a useful starting point for thinking through how the processes of assembling cities are driven in ways that are not completely reducible to macro structural forces.

8.4   Encounter and Futures

As we have seen throughout *Global Asian City*, migration is playing an important role in reconfiguring urban life in Seoul and altering the present position of the city vis-à-vis other territories. In focusing on notions of desire as generative of migration and its articulation with urban assemblages I have also sought to place emphasis on examining the *encounter* between individual migrants and urban life, on the effects of national migration regimes and the influence this has on the negotiation of everyday spaces. In this regard, there is also a contribution to be made here to scholarly understandings of encounter and diversity (Amin & Thrift 2002, 2013; Leitner 2012; Valentine 2008; Ye 2016b), particularly as they are situated in relation to the making of urban life.

The notion of encounter as it has been developed here does not simply refer to the meeting of two separate and distinct entities, migration and cities or people from different places. Rather, as I have emphasised throughout this book, migration and its relation to assembling urban life necessarily involves relational processes of becoming and change, for migrants, for the city and for the multiple relations that are constituted between them. Throughout Chapters 4, 5 and 6, we have seen that migration does not lead to effects that are anticipated in advance by people on the move, or indeed by governments, but rather takes shape in diverse processes of becoming (Papadopoulos & Tsianos 2008). Encounters with the rules and regulation of migration regimes, with workplaces and public spaces, with other migrant and non-migrant colleagues, peers and residents and with the possibilities that exist in life in the city have all been significant in the migrant lives traced in this book. These encounters have shaped what is possible for people arriving and living in Seoul, sometimes circumscribing lives to closed social networks and identity formations, sometimes involving confrontation or collaboration with others and at other times leading to shifts in subjectivity, recognising common connections with others and even in some instances aligning oneself with a sense of being Korean despite its ostensible untenability.

While it has been clear that there is no unified process of encounter and becoming, even amongst workers, teachers and students, there have been relatively common patterns that have emerged that reflect the circumstances of individuals who people these categories. There appear to be quite clear demarcations in terms of what is possible for differently embodied migrants: white and non-white, East Asian and Southeast Asian, male or female or workers and students. This would seem to support claims made by geographers about the significance of pre-existing ideas around cultural difference shaping the potential of encounters and also being reinscribed through those encounters (Valentine 2008; Leitner 2012). At the same time, the focus here on migration and urban life has also highlighted that these differences in encounters relate in important ways to the spatial and temporal locatedness of migrant lives. As noted above and throughout this book, workers, teachers and students live very different lives that reflect the roles they are accorded in the migration regime and the practices of employers and universities. Accordingly, they have different scope for encountering others and are empowered in different ways in those encounters. Recognising how encounters shape migrant lives and their contribution to the city demands attention to the uneven geography of migration and urban incorporation.

By working with notions of desire and urban assemblage it has also been possible to move beyond treatments of space as a passive backdrop for encounters and their effects. Much of the literature on encounters, particularly that focused on cultural difference, tends to emphasise the immediacy of the encounter and its effects on human participants (Amin & Thrift 2002, 2013; Valentine 2008; Ye 2016b). What we have seen in *Global Asian City* however is that migration and the encounters it develops reconstitute urban life in ways that disrupt and reconfigure existing understandings of social identity. Historically, this has been achieved through the development of urban subjectivities amongst Koreans who migrated to Seoul. In this book too we have seen how international students contribute to the constitution of diverse campus spaces and their cosmopolitan subjectivities; how workers disrupt the normative spaces of factories and make space for themselves as active urban agents; and how temporary urban spaces emerge around socialising activities and shared identities of different migrants. These manifestations of difference are not only a by-product of encounters but rather need to be understood as the very constitution of encounters and their effects in the social and material spaces of the city.

The analysis in this book has then drawn attention to the way in which encounters exceed the actual event of encounter (Wilson & Darling 2016). The focus on encounter, and the way in which it highlights transformations in urban and migrant lives, reveals futures that are not only limited to either being part of the present city or not. While specific encounters themselves have an immediate effect on individuals we have also seen that they contribute to the way those individuals orient themselves towards the city, taking initiative to cross boundaries of cultural difference or pursuing security amongst familiar but perhaps more closed

social networks. The notion of coupling and decoupling of migrant lives identified in Chapter 7 also points towards the ways that these encounters in the city
and the orientations that emerge contribute to the reconfiguration of future possibility – for some a place in the city, while for others closure of that opportunity
or the desire to become through circulation elsewhere. Seoul itself is also being
reconfigured through processes of migration in ways that will endure beyond the
presence of the individuals whose narratives have been foregrounded in this book.
Seoul's ongoing encounter with migration manifests a reconfiguration of the
urban assemblage, the emergence of new alliances and subjectivities, the transformation of local, national, regional and global territories and the re-imagination of what it might mean to be a *Global Asian City*.

8.5   For Other Approaches to Migration

In closing this book, I want to highlight the implications of the theoretical arguments developed here for current approaches to migration. As I have noted from
the outset of this book contemporary approaches to migration are guided by an
emphasis on migration management, an approach that seeks to generate what are
claimed to be 'mutually beneficial' outcomes for migrants and sending and
receiving states (Kalm 2010). As a political rationality migration management
depends on the assumption that migrants are utility-maximising individuals who
will make migration decisions that will serve their interests and for which they
can be held responsible. Migrants are accordingly classed as either voluntary,
such as those in this book, or involuntary/forced (Ottonelli & Torresi 2013).
Within the former, categories of graduated skill, gender, age and ethnicity are
also employed to situate migrants in differentiated positions in their migration
and in their everyday lives (Mezzadra & Neilson 2013). These assumptions make
it possible for policymakers to argue for systems of 'regulated openness' to
'balance the needs and interests of the sending, receiving and transit countries
and the migrants themselves' (Ghosh 2007: 107).

Mainstream migration studies have been an active contributor to the assumption of utility-maximisation and its hegemonic position in migration policymak-
ing. There has been a massive growth in the quantity of scholarship on migration
over the last decade, much of it undertaken not to advance new ways of thinking
about and understanding migration per se but rather to evaluate and assess the
outcomes of particular migration programmes (Garip 2012). While such research
claims to be politically neutral and even scientific in orientation it needs to be
recognised for the ways in which it contributes to maintaining current approaches
to managing migration and the lives of migrants (Hess 2010). The figure of
the utility-maximising migrant is the most pervasive component of this knowledge
(Nail 2015), although quantification of migrant populations, modelling of future
trends and evaluations of settlement outcomes can all be included as attempts to

'render knowledge of migration as an object of governmentality' (Casas-Cortes et al. 2015: 63). The categorisation of migrants as being either voluntary/involuntary, skilled/unskilled or from one ethnic group or another and then distinguishing them analytically as well as theoretically is another dimension of this knowledge of migration. It makes it possible to conceive of groups as inherently different and then to reinforce that through the development of distinct conceptualisations and approaches to their management.

The focus on desire in this book does not necessarily completely displace these knowledges and their presumed value for regulating populations but it does challenge the unthinking assumption about what drives migration and how migrants can or should be categorised analytically. Most notably, the claim that migrants make individual choices about whether to migrate sustains the argument that migration policies that differentially include and exclude migrants are appropriate. Read through the lens of desire, the generation of migration through policy, regulation and rules, the activities of intermediaries and the circulation of images about places must be understood as also part of an active attempt to diminish and differentiate migrant status in the social formations of the city and the nation. While we have seen that such efforts are never fully successful at the practical level they are pervasive in policy and popular discourses – the idea that migrants should be denied rights in a graduated way when they arrive in a new place (Anderson 2013; Bauder 2016). In places like Seoul and South Korea, and indeed globally, these rationalities justify long-term maintenance of non-citizenship for increasingly large sections of populations (Landolt & Goldring 2015). I argue that a focus on desire demands a reassessment of these approaches to migration and to the ethics of differentially including and excluding mobile people from the possibilities of social life.

References

Abdullah, N. (2005). Foreign bodies at work: good, docile and other-ed. *Asian Journal of Social Science*, 33(2), 223–245.

Agnew, J. A. (2003). *Geopolitics: Re-visioning World Politics*. Oxon: Psychology Press.

Ahmad, A.N. (2009). Bodies that (don't) matter: desire, eroticism and melancholia in Pakistani labour migration. *Mobilities*, 4(3), 309–327.

Ahmed, S. (2002). This other and other others. *Economy & Society*, 31(4), 558–572.

Ahmed, S., Castañeda, C. Fortier, A.M. & Sheller, M. (2003). *Uprootings/Regroundings: Questions of Home and Migration*. Oxford: Berg.

Ahn, H. (2014). Teachers' attitudes towards Korean English in South Korea. *World Englishes*, 33(2), 195–222.

Ahn, J.H. (2012). Transforming Korea into a multicultural society: reception of multiculturalism discourse and its discursive disposition in Korea. *Asian Ethnicity*, 13(1), 97–109.

Amin, A. (2002). Ethnicity and the multicultural city: living with diversity. *Environment & Planning A*, 34(6), 959–980.

Amin, A. (2013). Land of strangers. *Identities*, 20(1), 1–8.

Amin, A. & Thrift, N. (2002). *Cities: Reimagining the Urban*. Cambridge: Polity Press.

Amrith, S.S. (2011). *Migration and Diaspora in Modern Asia*. Cambridge: Cambridge University Press.

Anderson, B. (2009). What's in a name? Immigration controls and subjectivities: the case of au pairs and domestic worker visa holders in the UK. *Subjectivity*, 29(1), 407–424.

Anderson, B. (2010). Migration, immigration controls and the fashioning of precarious workers. *Work, Employment & Society*, 24(2), 300–317.

Anderson, B. (2013). *Us and Them?: The Dangerous Politics of Immigration Control*. Oxford: Oxford University Press.

Anderson, B. & McFarlane, C. (2011). Assemblage and geography. *Area*, 43(2), 124–127.

Andrucki, M.J. & Dickinson, J. (2015). Rethinking centers and margins in geography: bodies, life course, and the performance of transnational space. *Annals of the Association of American Geographers*, 105(1), 203–218.

Asis, M. & Piper, N. (2008). Researching international labor migration in Asia. *The Sociological Quarterly*, 49(3), 423–444.

Bailey, A. J., Wright, R.A., Mountz, A. & Miyares, I.M. (2002). (Re) producing Salvadoran transnational geographies. *Annals of the Association of American Geographers*, 92(1), 125–144.

Battistella, G. ed. (2014). *Global and Asian Perspectives on International Migration.* Heidelberg: Springer.

Bauder, H. (2016). *Migration Borders Freedom.* London: Routledge.

Berlant, L.G. (2011). *Cruel Optimism.* Duke University Press.

Bignall, S. (2008). Deleuze and Foucault on desire and power. *Angelaki: Journal of Theoretical Humanities*, 13(1), 127–147.

Bignall, S. & Patton, P. (2010). *Deleuze and the Postcolonial.* Edinburgh: Edinburgh University Press.

Bothwell, E. (2015). South Korea plans 'ghettoised' university courses for foreign students. *Times Higher Education Supplement.* August 14.

Botterill, K. (2016). Discordant lifestyle mobilities in East Asia: privilege and precarity of British retirement in Thailand. *Population, Space & Place.* doi: 10.1002/psp.2011.

Brady, M. (2014). Ethnographies of neoliberal governmentalities: from the neoliberal apparatus to neoliberalism and governmental assemblages. *Foucault Studies*, (18), 11–33.

Brenner, N., Madden, D.J. & Wachsmuth, D. (2011). Assemblage urbanism and the challenges of critical urban theory. *City*, 15(2), 225–240.

Brooks, R. & Waters, J. (2011). *Student Mobilities, Migration and the Internationalization of Higher Education.* Houndmills: Palgrave Macmillan.

Butler, J. (1987). *Subjects of Desire: Hegelian Reflections in Twentieth-Century France.* New York: Columbia University Press.

Carling, J.R. (2002). Migration in the age of involuntary immobility: theoretical reflections and Cape Verdean experiences. *Journal of Ethnic & Migration Studies*, 28(1), 5–42.

Carling, J. & Schewel, K. (2017). Revisiting aspiration and ability in international migration. *Journal of Ethnic & Migration Studies*, 1–19.

Casas-Cortes, M., Cobarrubias, S., De Genova, N. et al. (2015). New keywords: migration and borders. *Cultural Studies*, 29(1), 55–87.

Castles, S. (1995). How nation-states respond to immigration and ethnic diversity. *Journal of Ethnic & Migration Studies*, 21(3), 293–308.

Castles, S. (2010). Understanding global migration: a social transformation perspective. *Journal of Ethnic & Migration Studies*, 36(10), 1565–1586.

Chang, K.S. (1999). Compressed modernity and its discontents: South Korean society in transition. *Economy & Society*, 28(1), 30–55.

Cheng, S. (2011). *On the Move for Love: Migrant Entertainers and the US Military in South Korea.* University of Pennsylvania Press.

Cho, M.R. (1997). Flexibilization through metropolis: the case of postfordist Seoul, Korea. *International Journal of Urban & Regional Research*, 21(2), 180–201.

Cho, Y.H. & Palmer, J.D. (2013). Stakeholders' views of South Korea's higher education internationalization policy. *Higher Education*, 65(3), 291–308.

Choy, C.C. (2003). *Empire of Care: Nursing and Migration in Filipino American History*. Duke University Press.

Christou, A. (2009). Telling diaspora stories: theoretical and methodological reflections on narratives of migrancy and belongingness in the second generation. *Migration Letters*, 6(2), 143–153.

Chu, J.Y. (2010). *Cosmologies of Credit: Transnational Mobility and the Politics of Destination in China*. Duke University Press.

Collins, F.L. (2009). Transnationalism unbound: detailing new subjects, registers and spatialities of cross-border lives. *Geography Compass*, 3(1), 434–458.

Collins, F.L. (2010). International students as urban agents: international education and urban transformation in Auckland, New Zealand. *Geoforum*, 41(6), 940–950.

Collins, F.L. (2012). Transnational mobilities and urban spatialities. Notes from the Asia-Pacific. *Progress in Human Geography*, 36(3), 316–335.

Collins, F. (2014a). Globalising higher education in and through urban spaces: higher education projects, international student mobilities and trans-local connections in Seoul. *Asia Pacific Viewpoint*, 55(2), 242–257.

Collins, F.L. (2014b). Teaching English in South Korea: mobility norms and higher education outcomes in youth migration. *Children's Geographies*, 12(1), 40–55.

Collins, F.L. (2016). Labour and life in the global Asian city: the discrepant mobilities of migrant workers and English teachers in Seoul. *Journal of Ethnic & Migration Studies*, 42(14), 2309–2327.

Collins, F.L. (2017). Desire as a theory for migration studies: temporality, assemblage and becoming in the narratives of migrants. *Journal of Ethnic & Migration Studies*. doi: 10.1080/1369183X.2017.1384147

Collins, F.L. & Ho, K.C. (2014). Globalising higher education and cities in Asia and the Pacific. *Asia Pacific Viewpoint*, 55(2), 127–131.

Collins, F.L., Ho, K., Ishikawa, M. & Ma, A.H.S. (2017). International student mobility and after-study lives: the portability and prospects of overseas education in Asia. *Population, Space & Place*, 23(4): e2029.

Collins, F.L. & Huang, S. (2012). Migration methodologies: emerging subjects, registers and spatialities of migration in Asia. *Area*, 44(3), 270–273.

Collins, F.L., Lai, A.E. & Yeoh, B.S.A. (2013). Approaching Migration and Diversity in Asian Contexts. In A.E. Lai, F.L. Collins & B.S.A. Yeoh (eds.), *Migration and Diversity in Asian Contexts*, Singapore: Institute of South East Asian Studies Publishing, pp. 1–28.

Collins, F.L. & Pak, S. (2008). Language and skilled migration: the outcomes of overseas language study for South Korean students in New Zealand. *Asian Population Studies*, 4(3), 347–362.

Collins, F.L. & Park, G.S. (2016). Ranking and the multiplication of reputation: reflections from the frontier of globalizing higher education. *Higher Education*, 72(1), 115–129.

Collins, F.L. & Shubin, S. (2015). Migrant times beyond the life course: the temporalities of foreign English teachers in South Korea. *Geoforum*, 62, 96–104.

Collins, F.L., Sidhu, R., Lewis, N. & Yeoh, B.S. (2014). Mobility and desire: international students and Asian regionalism in aspirational Singapore. *Discourse: Studies in the Cultural Politics of Education*, 35(5), 661–676.

Collyer, M. & King, R. (2015). Producing transnational space: international migration and the extra-territorial reach of state power. *Progress in Human Geography*, 39(2), 185–204.

Conradson, D. & McKay, D. (2007). Translocal subjectivities: mobility, connection, emotion. *Mobilities*, 2(2), 167–174.

Constable, N. (2009). Migrant workers and the many states of protest in Hong Kong. *Critical Asian Studies*, 41(1), 143–164.

Cranston, S. (2017). Calculating the migration industries: knowing the successful expatriate in the global mobility industry. *Journal of Ethnic & Migration Studies*, doi: 10.1080/1369183X.2017.1315517

Cresswell, T. (2010). Towards a politics of mobility. *Environment & Planning D: Society & Space*, 28(1), 17–31.

Darling, J. & Wilson, H.F. eds. (2016). *Encountering the City: Urban Encounters from Accra to New York*. Routledge.

Dauvergne, C. (2008). *Making People Illegal: What Globalization Means for Migration and Law*. Cambridge: Cambridge University Press.

Dean, M. (2009). *Governmentality: Power and Rule in Modern Society*. London: Sage.

de Dios, A. (2016). Packaging talent: the migrant creative labor management of overseas Filipino musicians. In *International Migration in Southeast Asia*. Singapore: Springer, pp. 181–209.

De Haas, H. (2011). *The Determinants of International Migration*. Working Paper 32, Oxford: International Migration Institute.

DeLanda, M. (2006). *A New Philosophy of Society: Assemblage Theory and Social Complexity*. London: Continuum.

Deleuze, G. & Guattari, F. (1983). *Anti-Oedipus: Capitalism and Schizophrenia*, Minneapolis: University of Minnesota Press.

Deleuze, G. & Guattari, F. (1986). *Kafka: Toward a Minor Literature*. Minneapolis: University of Minnesota Press.

Deleuze, G. & Guattari, F. (1987). *A Thousand Plateaus: Capitalism and Schizophrenia*. Minneapolis: University of Minnesota Press.

Deleuze, G. & Parnet, C. (2007). *Dialogues II*. New York: Columbia University Press.

Doty, R.L. (2003). *Anti-Immigrantism in Western Democracies: Statecraft, desire and the Politics of Exclusion*. New York: Routledge.

Doucette, J. (2013). Minjung tactics in a post-minjung era? The survival of self-immolation and traumatic forms of labour protest in South Korea. In G. Gall (ed.), *New Forms and Expressions of Conflict at Work*. Houndmills: Palgrave, pp. 212–230.

Dovey, K. (2011). Uprooting critical urbanism. *City*, 15(3–4), 347–354.

Edensor, T. (2011). Entangled agencies, material networks and repair in a building assemblage: the mutable stone of St Ann's Church, Manchester. *Transactions of the Institute of British Geographers*, 36(2), 238–252.

Elden, S. (2005). Missing the point: globalization, deterritorialization and the space of the world. *Transactions of the Institute of British Geographers*, 30(1), 8–19.

Elsheshtawy, Y. (2008). Transitory sites: mapping Dubai's 'forgotten' urban spaces. *International Journal of Urban & Regional Research*, 32(4), 968–988.

Erdal, M. & Oeppen, C. (2017). Forced to leave? The discursive and analytical significance of describing migration as forced and voluntary, *Journal of Ethnic & Migration Studies*. doi: 10.1080/1369183X.2017.1384149

Ettlinger, N. (2007). Precarity unbound. *Alternatives: Global, Local, Political*, 32(3), 319–340.

Farias, I. & Bender, T. eds. (2012). *Urban Assemblages: How Actor-Network Theory Changes Urban Studies*. London: Routledge.

Fechter, A.M. & Walsh, K. (2010). Examining 'expatriate' continuities: postcolonial approaches to mobile professionals. *Journal of Ethnic & Migration Studies*, 36(8), 1197–1210.

Fincher, R. & Shaw, K. (2009). The unintended segregation of transnational students in central Melbourne. *Environment & Planning A*, 41(8), 1884–1902.

Findlay, A.M., King, R., Smith, F.M. et al. (2012). World class? An investigation of globalisation, difference and international student mobility. *Transactions of the Institute of British Geographers*, 37(1), 118–131.

Fong, V. (2011). *Paradise Redefined: Transnational Chinese Students and the Quest for Flexible Citizenship in the Developed World*. Stanford: Stanford University Press.

Ford, M. & Piper, N. (2007). Southern sites of female agency: informal regimes and female migrant labour resistance in East and Southeast Asia. In J. Hobson & L. Seabrooke (eds.), *Everyday Politics of the World Economy*. Cambridge: Cambridge University Press, pp. 63–79.

Foucault, M. (1991) Governmentality, In G. Burchell, C. Gordon & P. Miller (eds) *The Foucault Effect*. Chicago: University of Chicago Press.

Freeman, C. (2011). *Making and Faking Kinship: Marriage and Labor Migration between China and South Korea*. New York: Cornell University Press.

Furlong, A. (2014). Young people and the social consequences of the post-industrial economy. In P. Kelly & A. Kamp (eds.) *A Critical Youth Studies for the 21st Century*. Leiden: Brill, pp. 25–37.

Garelli, G. & Tazzioli, M. (2013). Challenging the discipline of migration: militant research in migration studies, an introduction. *Postcolonial Studies*, 245–249.

Garip, F. (2012). Discovering diverse mechanisms of migration: the Mexico–US Stream 1970–2000. *Population & Development Review*, 38(3), 393–433.

Geiger, M. & Pécoud, A. eds. (2013). *Disciplining the Transnational Mobility of People*. Houndmills: Palgrave.

Gelézeau, V. (2008). Changing socio-economic environments, housing culture and new urban segregation in Seoul. *European Journal of East Asian Studies*, 7(2), 295–321.

Georgi, F. (2010). For the benefit of some: the international organization for migration and its global migration management. In M. Geiger & A. Pécoud (eds.), *The Politics of International Migration Management*. Houndmills: Palgrave Macmillan, pp. 45–72.

Ghorashi, H. (2010). From absolute invisibility to extreme visibility: emancipation trajectory of migrant women in the Netherlands. *Feminist Review*, 94(1), 75–92.

Ghosh, B. (2007). Managing migration: towards the missing regime? In A. Pécoud & P. de Guchteneire (eds.), *Migration Without Borders: Essays on the Free Movement of People*. Oxford: Berghahn, pp. 97–118.

Gill, N. (2016). *Nothing Personal?: Geographies of Governing and Activism in the British Asylum System*. John Wiley & Sons.

Glick Schiller, N. & Çağlar, A. (2009). Towards a comparative theory of locality in migration studies: migrant incorporation and city scale. *Journal of Ethnic & Migration Studies*, 35(2), 177–202.

Glick Schiller, N. & Çağlar, A. (2011). *Locating Migration: Rescaling Cities and Migrants*. New York: Cornell University Press.

Glick Schiller, N. & Salazar, N.B. (2013). Regimes of mobility across the globe. *Journal of Ethnic & Migration Studies*, 39(2), 183–200.

Goddard, J. & Vallance, P. (2013). *The University and the City*. Routledge.

Gogia, N. (2006). Unpacking corporeal mobilities: the global voyages of labour and leisure. *Environment & Planning A*, 38(2), 359–375.

Gray, K. (2007). From human to workers' rights: the emergence of a migrant workers' union movement in Korea. *Global Society*, 21(2), 297–315.

Griffiths, M.B. (2014). Out of time: the temporal uncertainties of refused asylum seekers and immigration detainees. *Journal of Ethnic & Migration Studies*, 40(12), 1991–2009.

Grossberg, L. (1993). Cultural studies and/in new worlds. In C. McCarthy (ed.), *Race, Identity, and Representation in Education*, New York: Routledge, pp. 89–105.

Grosz, E.A. (1994). *Volatile Bodies: Toward a Corporeal Feminism*. Bloomington: Indiana University Press.

Grzymala-Kazlowska, A. & Phillimore, J. (2017). Introduction: rethinking integration. New perspectives on adaptation and settlement in the era of super-diversity. *Journal of Ethnic & Migration Studies*. doi: 10.1080/1369183X.2017.1341706

Hall, S. (1993). Culture, community, nation. *Cultural Studies*, 7(3), 349–363.

Hall, S.M. (2015). Migrant urbanisms: ordinary cities and everyday resistance. *Sociology*, 49(5), 853–869.

Hall, S., King, J. & Finlay, R. (2016). Migrant infrastructure: transaction economies in Birmingham and Leicester, UK. *Urban Studies*, https://doi.org/10.1177/0042098016634586

Han, G.S. (2007). Multicultural Korea: celebration or challenge of multiethnic shift in contemporary Korea? *Korea Journal*, 47(4), 32–63.

Han, G.S. (2015). *Nouveau-riche Nationalism and Multiculturalism in Korea: A Media Narrative Analysis*. London: Routledge.

Han, K.K. (2007). The archaeology of the ethnically homogeneous nation-state and multiculturalism in Korea. *Korea Journal*, 47(1), 8–31.

Harkness, N. (2013). Softer soju in South Korea. *Anthropological Theory*, 13(1–2), 12–30.

Hess, S. (2010). 'We are facilitating states!' An ethnographic analysis of the ICMPD. In M. Geiger & A. Pécoud (eds.), *The Politics of International Migration Management*. Houndmills: Palgrave, pp. 96–118.

Hindman, H. & Oppenheim, R. (2014). Lines of labor and desire: 'Korean Quality' in contemporary Kathmandu. *Anthropological Quarterly*, 87(2), 465–495.

Ho, E.L.E. (2014). The emotional economy of migration driving Mainland Chinese transnational sojourning across migration regimes. *Environment & Planning A*, 46(9), 2212–2227.

Hoang, K.K. (2015). *Dealing in Desire: Asian Ascendancy, Western Decline, and the Hidden Currencies of Global Sex Work*. University of California Press.

Hoang, L.A. (2017). Governmentality in Asian migration regimes: the case of labour migration from Vietnam to Taiwan. *Population, Space & Place*, 23(3), e2019.

Horgan, M. & Liinamaa, S. (2017). The social quarantining of migrant labour: everyday effects of temporary foreign worker regulation in Canada. *Journal of Ethnic & Migration Studies*, 43(5), 713–730.

Hugo, G.J. (2006). Immigration responses to global change in Asia: a review. *Geographical Research*, 44(2), 155–172.

International Labour Organization (2010). *Pioneering a system of migration management in Asia: The Republic of Korea's Employment Permit System*. Geneva: International Labour Organization.

Isin, E.F. (2007). City. state: critique of scalar thought. *Citizenship Studies*, 11(2), 211–228.

Jeon, M. (2009). Globalization and native English speakers in English Programme in Korea (EPIK). *Language, Culture & Curriculum*, 22(3), 231–243.

Jo, Y.N. (2015). Disclosing the poverty–shame nexus within popular films in South Korea (1975–2010), In E. Chase & G. Bantebya-Kyomuhendo (eds), *Poverty and Shame: Global Experiences*. Oxford: Oxford University Press, pp. 86–98.

Jones, D. (2013). Cosmopolitans and 'cliques': everyday socialisation amongst Tamil student and young professional migrants to the UK. *Ethnicities*, 13(4), 420–437.

Jun, M.J., Ha, S.K. & Jeong, J.E. (2013). Spatial concentrations of Korean Chinese and determinants of their residential location choices in Seoul. *Habitat International*, 40, 42–50.

Kalir, B., Sur, M. & Schendel, W. (2012). *Transnational Flows and Permissive Polities: Ethnographies of Human Mobilities in Asia*. Amsterdam: Amsterdam University Press.

Kalm, S. (2010). Liberalizing movements? The political rationality of global migration management. In M. Geiger & A. Pécoud (eds.), *The Politics of International Migration Management*. Houndmills: Palgrave, pp. 21–44.

Kang, Y. (2012). Singlish or globish: multiple language ideologies and global identities among Korean educational migrants in Singapore. *Journal of Sociolinguistics*, 16(2), 165–183.

Karaman, O. (2012). An immanentist approach to the urban. *Antipode*, 44(4), 1287–1306.

Kennedy, M., Rutherford-Johnson, T. & Kennedy, J. (2013). *The Oxford Dictionary of Music*. Oxford: Oxford University Press.

Kim, A.E. (2009). Global migration and South Korea: foreign workers, foreign brides and the making of a multicultural society. *Ethnic & Racial Studies*, 32(1), 70–92.

Kim, D. (2011). Promoting migrants' rights in South Korea: NGOs and the enactment of the employment permit system. *Asian & Pacific Migration Journal*, 20(1), 55–78.

Kim, E. (2004). Itaewon as an alien space within the nation-state and a place in the globalization era. *Korea Journal*, 44(3), 34–64.

Kim, E.H. (2008). Keeping the gateway shut: regulating global city-ness in Seoul. In M. Price & L. Benton-Short (eds), *Migrants to the Metropolis: The Rise of Immigrant Gateway Cities*, Syracuse: Syracuse University Press, pp. 322–344.

Kim, J.K. (2011). The politics of culture in multicultural Korea. *Journal of Ethnic & Migration Studies*, 37(10), 1583–1604.

Kim, J.K. (2015). Embedded solidarity: International migrant labor advocacy in South Korea. In *Immigration and Work* (pp. 75–98). Emerald Group Publishing Limited.

Kim, M. (2013). Citizenship projects for marriage migrants in South Korea: intersecting motherhood with ethnicity and class. *Social Politics: International Studies in Gender, State & Society*, 20(4), 455–481.

Kim, N.H.J. (2012). Multiculturalism and the politics of belonging: the puzzle of multiculturalism in South Korea. *Citizenship Studies*, 16(1), 103–117.

Kim, S. (2009). Politics of representation in the era of globalization: discourse about marriage migrant women in two South Korean films. *Asian Journal of Communication*, 19(2), 210–226.

Kim, S. (2012). Racism in the global era: analysis of Korean media discourse around migrants, 1990–2009. *Discourse & Society*, 23(6), 657–678.

Kim, S.K. (2013). Framing the globalization debate in Korean higher education. *Korea 2013: Politics, Economy & Society*, 137–159

Kim, S.S. ed. (2000). *Korea's Globalization*. Cambridge: Cambridge University Press.

Kim, W.B. (1999). Developmentalism and beyond: reflections on Korean cities. *Korea Journal*. 39(3), 5–34.

Kim, W.B. (2004). Migration of foreign workers into South Korea: from periphery to semi-periphery in the global labor market, *Asian Survey*, 44(2), 316–335.

Kim, Y. (2013). *Transnational Migration, Media and Identity of Asian Women: Diasporic Daughters*. London: Routledge.

Kim, J. & Choe, S.C. (1997). *Seoul: The Making of a Metropolis*. Oxford: John Wiley & Sons.

Kim, J. & Hong, S. (2007). Queer cultural movements and local counterpublics of sexuality: a case of Seoul Queer Films and Videos Festival. *Inter-Asia Cultural Studies*, 8(4), 617–633.

Kim Watson, J. (2011). *The New Asian City: Three-Dimensional Fictions of Space and Urban Form*. Minneapolis: University of Minnesota Press.

Kitiarsa, P. (2008). Thai migrants in Singapore: state, intimacy and desire. *Gender, Place & Culture*, 15(6), 595–610.

KoILaF. (2007). Review on implementation of employment permit system. *Korea Labor Review*, 16, 2–10.

Kono, Y. & Chang, L. (2014). Trends and strategies for attracting international students to US public health programs. *World Education News & Reviews*, 27(3). April (2014). Available at: https://ssrn.com/abstract=2434534 [accessed 21 Dec 2017].

Koo, H. (2007). The changing faces of inequality in South Korea in the age of globalization. *Korean Studies*, 31(1), 1–18.

Korea Immigration Service (2017). *Statistics (2016)*. Seoul: Ministry of Justice.

Križnik, B. (2011). Selling global Seoul: competitive urban policy and symbolic reconstruction of cities. *Revija za sociologiju*, 3, 291–313.

Lai, A.E., Collins, F.L. & Yeoh, B.S.A. (2013). *Migration and Diversity in Asian Contexts*. Singapore: Institute of South East Asian Studies.

Lan, P.C. (2011). White privilege, language capital and cultural ghettoisation: western high-skilled migrants in Taiwan. *Journal of Ethnic & Migration Studies*, 37(10), 1669–1693.

Landolt, P. & Goldring, L. (2015). Assembling noncitizenship through the work of conditionality. *Citizenship Studies*, 19(8), 853–869.

Lartigue, C.J. (2000). You'll never guess what South Korea frowns upon. *Washington Post*, A1.

Lave J. & Wenger E. (1991). *Situated Learning: Legitimate Peripheral Participation*. Cambridge: Cambridge University Press.

Law, L. (2002). Defying disappearance: cosmopolitan public spaces in Hong Kong. *Urban Studies*, 39(9), 1625–1645.

Lawson, V.A. (2000). Arguments within geographies of movement: the theoretical potential of migrants' stories. *Progress in Human Geography*, 24(2), 173–189.

Lee, H.K. (2008). International marriage and the state in South Korea: focusing on governmental policy. *Citizenship Studies*, 12(1), 107–123.

Lee, H.Y. (2012). At the crossroads of migrant workers, class, and media: a case study of a migrant workers' television project. *Media, Culture & Society*, 34(3), 312–327.

Lee, O-Y. (1994). Seoul: an enormous piece of wrapping cloth. *Korea Journal*, 34(3): 69–72.

Leitner, H. (2012). Spaces of encounters: immigration, race, class, and the politics of belonging in small-town America. *Annals of the Association of American Geographers*, 102(4), 828–846.

Lewis, H., Dwyer, P., Hodkinson, S. & Waite, L. (2015). Hyper-precarious lives. Migrants, work and forced labour in the Global North. *Progress in Human Geography*, 39(5), 580–600.

Ley, D. (2010). *Millionaire Migrants: Trans-Pacific Life Lines*. Oxford: John Wiley & Sons.

Li, W. (1998). Anatomy of a new ethnic settlement: the Chinese ethnoburb in Los Angeles. *Urban Studies*, 35(3), 479–501.

Lie, J. (2014). *Multiethnic Korea?: Multiculturalism, Migration, and Peoplehood Diversity in Contemporary South Korea*. Institute of East Asian Studies, University of California, Berkeley.

Lim, T.C. (2002). The changing face of South Korea: the emergence of Korea as a 'land of immigration'. *The Korea Society Quarterly*, 3(2&3), 16–21.

Lim, T.C. (2003). Racing from the bottom in South Korea?: the nexus between civil society and transnational migrants. *Asian Survey*, 43(3), 423–442.

Lim, T. (2010). Rethinking belongingness in Korea: transnational migration, 'migrant marriages' and the politics of multiculturalism. *Pacific Affairs*, 83(1), 51–71.

Lim, T.C. (2014). Late migration, discourse, and the politics of multiculturalism in South Korea: a comparative perspective. In J. Lie (ed.) *Multiculturalism, Migration, and Peoplehood Diversity in Contemporary South Korea*. Institute of East Asian Studies: Berkeley, pp. 31–57.

Lin, W., Lindquist, J., Xiang, B. & Yeoh, B.S. (2017). Migration infrastructures and the production of migrant mobilities. *Mobilities*, 12(2), 167–174.

Lindquist, J. (2009). *The Anxieties of Mobility: Migration and Tourism in the Indonesian Borderlands*. Honolulu: University of Hawaii Press.

Lindquist, J. (2015). Of figures and types: brokering knowledge and migration in Indonesia and beyond. *Journal of the Royal Anthropological Institute*, 21(S1), 162–177.

Lindquist, J., Xiang, B. & Yeoh, B.S. (2012). Opening the black box of migration: brokers, the organization of transnational mobility and the changing political economy in Asia. *Pacific Affairs*, 85(1), 7–19.

Ling, L.H.M. (1984). East Asian migration to the Middle East causes, consequences and considerations. *International Migration Review*, 18(1), 19–36.

Lukacs, G. (2015). Labor games: youth, work, and politics in East Asia. *Positions: Asia Critique*, 23(3), 381–409.

Lundström, C. (2014). *White Migrations: Gender, Whiteness and Privilege in Transnational Migration*. Houndmills: Palgrave.

Luxner, A. (2004). *American English: A Teacher's Journey in Seoul, South Korea*, USA: Lightning Source Incorporated.

Marcu, S. (2017). Tears of time: a Lefebvrian rhythmanalysis approach to explore the mobility experiences of young Eastern Europeans in Spain. *Transactions of the Institute of British Geographers*, 42(3), 405–416.

Marrati, P. (2006). Time and affects: Deleuze on gender and sexual difference. *Australian Feminist Studies*, 21(51), 313–325.

McCann, E. & Ward, K. (2012). Assembling urbanism: following policies and 'studying through' the sites and situations of policy making. *Environment & Planning A*, 44(1), 42–51.

McCormack, D.P. & Schwanen, T. (2011). Guest editorial: the space-times of decision making. *Environment & Planning A*, 43, 2801–2818.

McFarlane, C. (2011a). The city as assemblage: dwelling and urban space. *Environment & Planning D*, 29(4), 649–671.

McFarlane, C. (2011b). *Learning the City: Knowledge and Translocal Assemblage*. Oxford: John Wiley & Sons.

McGee, T.G. (1971). *The Urbanization Process in The Third World*. London: G. Bell and Sons, Ltd.

McKay, D. (2007). 'Sending dollars shows feeling' – emotions and economies in Filipino migration. *Mobilities*, 2(2), 175–194.

Mezzadra, S. & Neilson, B. (2012). Between inclusion and exclusion: on the topology of global space and borders. *Theory, Culture & Society*, 29(4–5), 58–75.

Mezzadra, S. & Neilson, B. (2013). *Border as Method, or, the Multiplication of Labor*. Durham: Duke University Press.

Mills, M.B. (1999). *Thai Women in the Global Labor Force: Consuming Desires, Contested Selves*. Rutgers University Press.

Milne, R.S. (1986). Malaysia-beyond the new economic policy. *Asian Survey*, 26(12), 1364–1382.

Min, K.S. (2011). The progress and future orientation of the employment permit system (EPS). *Korea Labor Review*, 7(38), 2–6.

Mitchell, D. (1996). *The Lie of the Land: Migrant Workers and the California Landscape*. Minneapolis: University of Minnesota Press.

Mitchell, K. (1993). Multiculturalism, or the united colors of capitalism?. *Antipode*, 25(4), 263–294.

Mitchell, K. (2004). *Crossing the Neoliberal Line: Pacific Rim Migration and the Metropolis*. Philadelphia: Temple University Press.

Moon, K.H. (2000). Strangers in the midst of globalization: migrant workers and Korean nationalism. In S. Kim (ed.), *Korea's Globalization*, New York: Columbia University Press, pp. 147–169.

Moon, R.J. (2016). Internationalisation without cultural diversity? Higher education in Korea. *Comparative Education*, 52(1), 91–108.

Nail, T. (2015). *The Figure of the Migrant*. Stanford: Stanford University Press.

NIIED. (2016). *English Program in Korea*. National Institute of International Education and Development. Available at: https://www.epik.go.kr:8080/index.do [accessed 08 December 2015].

Nuttall, S. & Mbembe, A. (2008). *Johannesburg: The Elusive Metropolis*. Durham: Duke University Press.

OECD. (2012). *Education at a Glance (2012)*. Paris: OECD Publishing.

Ottonelli, V. & Torresi, T. (2013). When is migration voluntary? *International Migration Review*, 47(4), 783–813.

Ong, A. (2005). Ecologies of expertise: assembling flows, managing citizenship. In A. Ong & S. Collier (eds.) *Global Assemblages: Technology, Politics, and Ethics as Anthropological Problems*, Oxford: Blackwell, pp. 337–353.

Ong, A. (2007). Please stay: pied-a-terre subjects in the megacity. *Citizenship Studies*, 11(1), 83–93.

Papadopoulos, D., Stephenson, N. & Tsianos, V. (2008). *Escape Routes: Control and Subversion in the Twenty-First Century*. London: Pluto Press.

Papadopoulos, D. & Tsianos, V. (2008). The autonomy of migration: the animals of undocumented mobility. In A. Hickey-Moody & P. Malins (eds.), *Deleuzian Encounters: Studies in Contemporary Social Issues*. Houndmills: Palgrave-Macmillan.

Papadopoulos, D. & Tsianos, V.S. (2013). After citizenship: autonomy of migration, organisational ontology and mobile commons. *Citizenship Studies*, 17(2), 178–196.

Park, H.O. (2015). *The Capitalist Unconscious: From Korean Unification to Transnational Korea*. New York: Columbia University Press.

Park, J.K. (2009). 'English fever' in South Korea: its history and symptoms. *English Today*, 25(1), 50–57.

Park, J.S.Y. (2009). *The Local Construction of a Global Language: Ideologies of English in South Korea*. Walter de Gruyter.

Park, R.E., Burgess, E.W. & McKenzie, R.D. (1925). *The City*. University of Chicago Press.

Park, W.W. (2002). The unwilling hosts: state, society and the control of guest workers in South Korea. *Asia Pacific Business Review*, 8(4), 67–94.

Pécoud, A. (2014). *Depoliticising Migration: Global Governance and International Migration Narratives*. Palgrave.

Phan, L.P. (2016). *Transnational Education Crossing 'Asia' and 'the West'*. New York: Routledge.

Piper, N. (2006). Gendering the politics of migration. *International Migration Review*, 40(1), 133–164.

Platt, M., Baey, G., Yeoh, B.S. et al. (2017). Debt, precarity and gender: male and female temporary labour migrants in Singapore. *Journal of Ethnic & Migration Studies*, 43(1), 119–136.

Portes, A. & Jensen, L. (1989). The enclave and the entrants: patterns of ethnic enterprise in Miami before and after Mariel. *American Sociological Review*, 929–949.

Pratt, M.L. (1992). *Imperial Eyes: Travel Writing and Transculturation*. New York: Routledge.

Price, M. & Benton-Short, L. eds. (2008). *Migrants to the Metropolis: The Rise of Immigrant Gateway Cities*. Syracuse: Syracuse University Press.

Raghuram, P. (2013). Theorising the spaces of student migration. *Population, Space & Place*, 19(2), 138–154.

Rajkumar, D., Berkowitz, L., Vosko, L.F. et al. (2012). At the temporary–permanent divide: how Canada produces temporariness and makes citizens through its security, work, and settlement policies. *Citizenship Studies*, 16(3–4), 483–510.

Rath, J. ed. (2007). *Tourism, Ethnic Diversity and the City*. Routledge.

Robertson, S. (2011). Cash cows, backdoor migrants, or activist citizens? International students, citizenship, and rights in Australia. *Ethnic & Racial Studies*, 34(12), 2192–2211.

Robertson, S. (2013). *Transnational Student-Migrants and the State: The Education–Migration Nexus*. Houndmills: Palgrave Macmillan.

Robertson, S. (2014). Time and temporary migration: the case of temporary graduate workers and working holiday makers in Australia. *Journal of Ethnic & Migration Studies*, 40(12), 1915–1933.

Robinson, J. (2011). Cities in a world of cities: the comparative gesture. *International Journal of Urban & Regional Research*, 35(1), 1–23.

Rodriguez, R.M. & Schwenken, H. (2013). Becoming a migrant at home: subjectivation processes in migrant-sending countries prior to departure. *Population, Space & Place*, 19(4), 375–388.

Rogaly, B. (2015). Disrupting migration stories: reading life histories through the lens of mobility and fixity. *Environment & Planning D: Society and Space*, 33(3), 528–544.

Rose, N. (1999). *Powers of Freedom: Reframing Political Thought*. Cambridge: Cambridge University Press.

Roy, A. (2011). Slumdog cities: rethinking subaltern urbanism. *International Journal of Urban & Regional Research*, 35(2), 223–238.

Rubin, A. (2012). *Archives of Authority: Empire, Culture, and the Cold War*. Princeton: Princeton University Press.

Rubinov, I. (2014). Migrant assemblages: building postsocialist households with Kyrgyz remittances. *Anthropological Quarterly*, 87(1), 183–215.

Said, E.W. (1993). *Culture and Imperialism*. London: Vintage.

Salazar, N.B. (2011). The power of imagination in transnational mobilities. *Identities*, 18(6), 576–598.

Sassen, S. (2001). *The Global City*. New York: Princeton University Press.

Schinkel, W. & Van Houdt, F. (2010). The double helix of cultural assimilationism and neo-liberalism: citizenship in contemporary governmentality. *The British Journal of Sociology*, 61(4), 696–715.

Seo, S. & Skelton, T. (2017). Regulatory migration regimes and the production of space: the case of Nepalese workers in South Korea. *Geoforum*, 78, 159–168.

Seol, D.H. (2000). Past and present of foreign workers in Korea, 1987–2000. *Asia Solidarity Quarterly*, 2, 6–31.

Seol, D.H. (2012). The citizenship of foreign workers in South Korea. *Citizenship Studies*, 16(1), 119–133.

Seol, D.H. & Skrentny, J.D. (2009). Why is there so little migrant settlement in East Asia? *International Migration Review*, 43(3), 578–620.

Seoul Metropolitan Government (2002). *Vision Seoul 2006: The 4-year Master Plan for Municipal Policies*. Seoul: Seoul Metropolitan Government.

Seoul Metropolitan Government (2008*). Soul of Asia-SEOUL: A City of Design and Culture*. Seoul: Seoul Metropolitan Government.

Shamir, R. (2005). Without borders? Notes on globalization as a mobility regime. *Sociological Theory*, 23(2), 197–217.

Shi, Y. & Collins, F.L. (2017). Producing mobility: visual narratives of the rural migrant worker in Chinese television. *Mobilities*, doi: 10.1080/17450101.2017.1320133.

Shim, D. (2006). Hybridity and the rise of Korean popular culture in Asia. *Media, Culture & Society*, 28(1), 25–44.

Shim, D. & Park, J.S. (2008). The language politics of 'English fever' in South Korea. *Korea Journal*, 48(2), 136–59.

Shin, G.W. (2006). *Ethnic Nationalism in Korea: Genealogy, Politics, and Legacy*. Stanford: Stanford University Press.

Shin, G.W. & Choi, J. N. (2015). *Global Talent: Skilled Labor as Social Capital in Korea*. Stanford: Stanford University Press.

Shin, H. (2007). English language teaching in Korea: toward globalization or glocalization. In J. Cummins & C. Davison (eds.), *International Handbook of English Language Teaching*. New York: Springer, pp. 75–86.

Shubin, S. (2012). Living on the move: mobility, religion and exclusion of Eastern European migrants in rural Scotland. *Population, Space & Place*, 18(5), 615–627.

Shubin, S. (2015). Migration timespaces: a Heideggerian approach to understanding the mobile being of Eastern Europeans in Scotland. *Transactions of the Institute of British Geographers*, 40(3), 350–361.

Sidhu, R., Collins, F., Lewis, N. & Yeoh, B. (2016). Governmental assemblages of internationalising universities: mediating circulation and containment in East Asia. *Environment & Planning A*, 48(8), 1493–1513.

Silvey, R. (2004). Power, difference and mobility: feminist advances in migration studies. *Progress in Human Geography*, 28(4), 490–506.

Simone, A. (2004). People as infrastructure: intersecting fragments in Johannesburg. *Public Culture*, 16(3), 407–429.

Simone, A.M. (2010). *City Life from Jakarta to Dakar*. New York: Routledge.

Simone, A. (2011). The surfacing of urban life: a response to Colin McFarlane and Neil Brenner, David Madden and David Wachsmuth. *City*, 15(3–4), 355–364.

Smith, D.W. (2007). Deleuze and the question of desire: toward an immanent theory of ethics. *Parrhesia*, 2, 66–78.

Smith, M.P. (2000). *Transnational Urbanism: Locating Globalization*. Malden: Blackwell.

Song, D. (2013). Spatial process and cultural territory of Islamic Food restaurants in Itaewon, Seoul. In A.E. Lai, F.L. Collins & B.S.A. Yeoh (eds.) *Migration and Diversity in Asian Contexts*, Singapore: Institute of South East Asian Studies Publishing, pp. 233–253.

Song, J. (2015). Labour markets, care regimes and foreign care worker policies in East Asia. *Social Policy & Administration*, 49(3), 376–393.

Specht, A. & Freeborne, J. (1996). *Korea Calling: The Essential Handbook for Teaching English and Living in South Korea*. USA: Woodpecker Press.

Sun, W. (2002). *Leaving China: Media, Migration, and Transnational Imagination*. Rowman & Littlefield Publishers.

Tazzioli, M. & Walters, W. (2016). The sight of migration: governmentality, visibility and Europe's contested borders. *Global Society*, 30(3), 445–464.

Tsing, A.I. (2005). *Friction: An Ethnography of Global Connection*. Princeton: Princeton University Press.

Tsing, A.L. (2015). *The Mushroom at the End of the World: On the Possibility of Life in Capitalist Ruins*. Princeton: Princeton University Press.

Tyner, J.A. (2013). *Made in the Philippines*. New York: Routledge.

United Nations (2015). *International Migration Report*. New York: United Nations.

Valentine, G. (2008). Living with difference: reflections on geographies of encounter. *Progress in Human Geography*, 32(3), 323–337.

Vertovec, S. (2007). Super-diversity and its implications. *Ethnic & Racial Studies*, 30(6), 1024–1054.

Wagner, B.K. & Van Volkenburg, M. (2011). HIV/AIDS tests as a proxy for racial discrimination-a preliminary investigation of South Korea's policy of mandatory in-country HIV/AIDS tests for its foreign English teachers. *Journal of Korean Law*, 11, 179–245.

Wang, C. (2013). Place of desire: skilled migration from mainland China to post-colonial Hong Kong. *Asia Pacific Viewpoint*, 54(3), 388–397.

Ward, K. (2010). Towards a relational comparative approach to the study of cities. *Progress in Human Geography*, 34(4), 471–487.

Waters, J.L. (2008). *Education, Migration, and Cultural Capital in the Chinese Diaspora.* Amherst: Cambria Press.

Waters, J.L. (2012). Geographies of international education: mobilities and the reproduction of social (dis) advantage. *Geography Compass,* 6(3), 123–136.

Watson, I. (2012). Cultural policy in South Korea: reinforcing homogeneity and cosmetic difference? *Journal of Asian Public Policy,* 5(1), 97–116.

Watson, I. (2013). (Re) constructing a world city: urbicide in global Korea. *Globalizations,* 10(2), 309–325.

Wilson, H.F. & Darling, J. (2016). The possibilities of encounter. In J. Darling & H.F. Wilson (eds.). *Encountering the City: Urban Encounters from Accra to New York.* Routledge.

Wise, A. (2005). Hope and belonging in a multicultural suburb. *Journal of Intercultural Studies,* 26(1–2), 171–186.

Wong, T.C. & Rigg, J. eds. (2010). *Asian Cities, Migrant Labor and contested Spaces.* Routledge.

Xiang, B. (2012). Predatory princes and princely peddlers: the state and international labour migration intermediaries in China. *Pacific Affairs,* 85(1), 47–68.

Xiang, B. & Lindquist, J. (2014). Migration infrastructure. *International Migration Review,* 48(1), 122–148.

Xiang, B. & Shen, W. (2009). International student migration and social stratification in China. *International Journal of Educational Development,* 29(5), 513–522.

Yang, P. (2016). *International Mobility and Educational Desire: Chinese Foreign Talent Students in Singapore.* Houndmills: Palgrave.

Ybiernas, V.A.S. (2013). Migrant workers in South Korea: between strategic ambivalence and systematic exploitation. *Social Science Diliman,* 9(1), 1–18.

Ye, J. (2014). Migrant masculinities: Bangladeshi men in Singapore's labour force. *Gender, Place & Culture,* 21(8), 1012–1028.

Ye, J. (2016a). *Class Inequality in the Global City: Migrants, Workers and Cosmopolitanism in Singapore.* Houndmills: Palgrave.

Ye, J. (2016b). Spatialising the politics of coexistence: gui ju (规矩) in Singapore. *Transactions of the Institute of British Geographers,* 41(1), 91–103.

Ye, J. (2017). Managing urban diversity through differential inclusion in Singapore. *Environment & Planning D: Society & Space,* doi: 10.1177/0263775817717988.

Yea, S. (2015). *Trafficking Women in Korea: Filipina Migrant Entertainers.* New York: Routledge.

Yeoh, B.S. (2004). Cosmopolitanism and its exclusions in Singapore. *Urban Studies,* 41(12), 2431–2445.

Yeoh, B.S. (2006). Bifurcated labour: the unequal incorporation of transmigrants in Singapore. *Tijdschrift voor Economische en Sociale Geografie,* 97(1), 26–37.

Yeoh, B.S. (2017). International migration and citizenship in Asia. *Migration & Citizenship,* 5(1), 7–13.

Yeoh, B.S. & Huang, S. (1998). Negotiating public space: strategies and styles of migrant female domestic workers in Singapore. *Urban Studies,* 35(3), 583–602.

Yeoh, B.S. & Huang, S. (2011). Introduction: fluidity and friction in talent migration. *Journal of Ethnic & Migration Studies,* 37(5), 681–690.

Yeoh, B.S. & Willis, K. (2005). Singaporean and British transmigrants in China and the cultural politics of 'contact zones'. *Journal of Ethnic & Migration Studies,* 31(2), 269–285.

Yi, T.J. (1995). The nature of Seoul's modern urban development during the 18th and 19th centuries. *Korea Journal*, 35(3), 5–30.

Yoon, S. (2014). The Qualifications for Being and Becoming English Language Teachers Across Junior/High School Level Public and Private Schools in Korea. Doctoral thesis, Indiana University of Pennsylvania.

Young, I.M. (1990). *Justice and the Politics of Difference*. Princeton: Princeton University Press.

Yun, J. (2011). Borderless village: challenging the globalist dystopia in Ansan, South Korea. *Traditional Dwellings & Settlements Review*, 22(2), 49–61.

Zhang, J., Lu, M.C.W. & Yeoh, B.S. (2015). Cross-border marriage, transgovernmental friction, and waiting. *Environment & Planning D: Society & Space*, 33, 229–246.

Index

accommodation *see* housing
age 6, 16, 26, 29, 31, 44, 132, 133, 176, 184, 192
Ansan 71, 80, 159, 161–3, 179
anticipatory politics 187
anxiety 13, 32, 35, 37, 62, 65, 66, 178, 179
Asian culture 62
Asian financial crisis (1997) 140
Asian Games (1986) 8, 9
assemblage
    assemblage thinking 39–40, 42, 103–4
    linking assemblage, desire and encounter 3, 13–16, 21, 37, 46, 125, 188–92
    national assemblages 41, 52, 154
    social assemblages 64, 119–21, 166, 175
    transnational assemblages 40–1
    urban assemblages 15–16, 41, 69, 74–7, 103–4, 106, 116, 139

belonging, sense of 123, 138–9, 147, 158, 172, 176–8, 188
biographical interviews 16–19
The 'borderless village' (Ansan) 159–60
borders 12, 21, 45
    border control in South Korea 49, 51–2, 69, 100

borders and migration control 24–8, 38, 74, 90, 103, 184
    internalisation of border practices 41, 53

Çağlar, Ayse 4–5, 7, 20, 182
Cheng, Sealing 33, 34–5, 37, 39, 102, 132, 134, 145
Cheonggyecheon stream 84, 153–4
citizenship 6, 7, 29, 38, 95, 99, 135
    non-citizenship 145, 193
The city
    city as bounded territory 39
    city as relationally constituted 19, 20, 39
class 6, 29, 30, 102, 129, 132, 148
Committee for Foreigner Policy 154
comparison 16, 19, 20, 41, 151
contact zones 44
contrapuntal approach 20–2, 48, 167
cosmopolitanism 16, 36, 42, 44, 105, 114, 117, 132, 167, 178, 191
cultural capital 10, 67, 68, 116, 132

debt 62, 85, 134, 135, 140, 143, 151, 183
Deleuze, Gilles 14–15, 27, 34, 35–8, 40, 175, 189
demographic renewal 48, 58, 102, 140, 183
deportation 76, 85, 98, 144

Global Asian City: Migration, Desire and the Politics of Encounter in 21st Century Seoul, First Edition. Francis L. Collins.
© 2018 John Wiley & Sons Ltd. Published 2018 by John Wiley & Sons Ltd.

subjectification 54, 74–5
subjectivity 29–30, 32–5, 37, 39, 74–5, 96,
    106, 128, 130–5, 139–40, 147, 150–2,
    155, 164, 166, 167, 171, 175, 179,
    190, 191–2

tactics of recognition 94–8
Test of Proficiency in Korean
    (TOPIK) 55
time and migration
    permanent temporariness 55,
        140, 144
    temporary migration 7, 25, 29, 30, 43,
        76, 99
    transience 51, 54, 89, 99–100, 131,
        139–44
    waiting 63, 65, 67, 68
Tsianos, Vassilis S. 22, 38, 73, 87, 142,
    177, 187, 190
Tsing, Anna 54, 74, 134, 135

urban periphery 71–9, 84–5, 91–4, 99,
    139, 151, 157, 159, 183, 184
urban space and migrant lives 5–6, 15,
    40–2, 77–85, 87, 153–4, 157–66

Walsh, Katie 133, 134
Waters, Johanna 58, 102, 105, 116
whiteness 22, 100, 126, 129–33, 151,
    161, 162
Wonggok-dong 159–60

Ye, Junjia 6, 7, 26, 29, 30, 41, 44, 45,
    77, 182
Yeoh, Brenda Saw Ai 7, 19, 26, 28–31, 37,
    41–2, 44, 54, 74, 156, 159, 184
youth
    South Korean youth 8, 12, 58, 70, 183
    youth and mobility 62, 64, 66–8, 132,
        134, 138, 141, 145
    youthfulness as desirable 58, 107, 132–3